D0409621

FATHER of LIONS

FATHER of LIONS

LOUISE CALLAGHAN

An Apollo Book

This is an Apollo book, first published in the UK in 2019 by
Head of Zeus Ltd

9 7 5 3 1 2 4 6 8

A catalogue record for this book is available from
the British Library.

ISBN (HB): 9781789540765
ISBN (XTPB): 9781789544541
ISBN (E): 9781789540789

Typeset by Divaddict Publishing Solutions Ltd

Maps © Jeff Edwards

Printed and bound in Great Britain by
CPI Group (UK) Ltd, Croydon CR0 4YY

Head of Zeus Ltd
First Floor East
5–8 Hardwick Street
London EC1R 4RG

WWW.HEADOFZEUS.COM

Contents

To MRM

Author's note

This book is a true story. The events that take place in it were described to me by the people who lived through them, although certain names have been changed. I cross-referenced their accounts and checked facts as closely as I could in what had recently been an active war zone. I spent hundreds of hours interviewing the main characters, in Mosul and elsewhere, trying to get as much detail as I could to bring this story to life.

A proportion of the proceeds from this book goes to the Four Paws charity, and to Abu Laith.

List of characters

Abu Laith – would-be zookeeper and Father of Lions
Lumia – Abu Laith's wife
Marwan – Abu Laith's spy inside the zoo

Abu Laith and Lumia's children and stepchildren (a selection):

Dalal – lives in Baghdad, works in military intelligence
Lubna, Oula, Mohammed – sent to Baghdad with Dalal
Luay – geography student
Abdulrahman – would-be zookeeper's apprentice
Nour, Ashraf (Geggo), Mo'men, Shuja

Hakam – chemist and guitar player
Hasna – Hakam's sister, student of English literature
Arwa – Hakam's mother
Said – Hakam's father
Abu Hareth – Isis member who despises Abu Laith
Ahmed – manager of the park where the zoo is located
Ibrahim – former owner of the animals
Muna – Abu Laith's first wife
Sara – Abu Laith's long-lost flame

Dr Amir – International rescue vet
Dr Suleyman – Kurdish vet
Marlies – Dr Amir's assistant
The Commander – a leader of the Iraqi army in Nineveh province
Heba – Marwan's fiancée

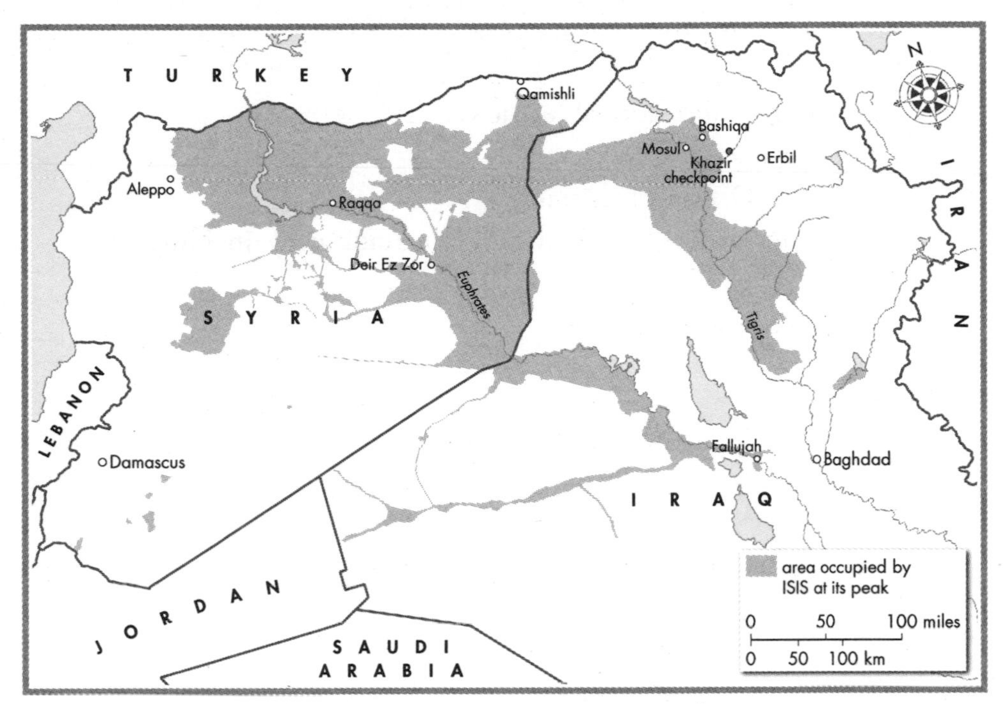

Source for information on ISIS territory: IHS Markit

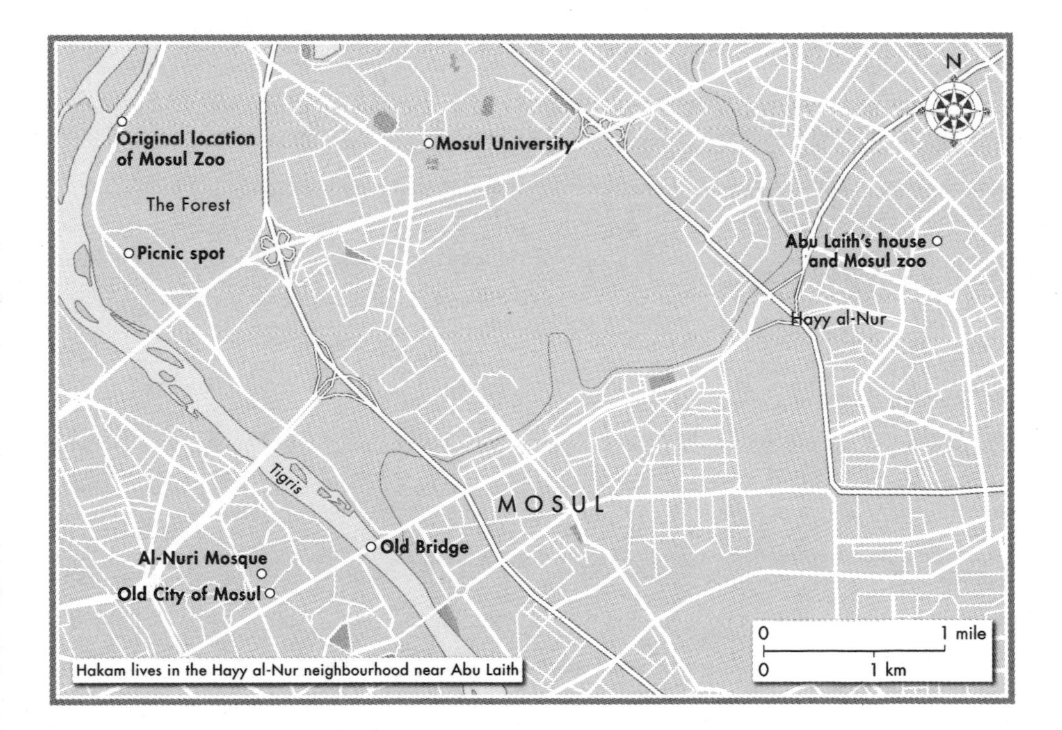

1

Abu Laith

Abu Laith was not the kind of man to let another man insult his lion. Especially not a man who looked like this.

He was wearing a short-sleeved shirt, well pressed, and had the air of a civil servant. He carried a baby in the crook of his left arm. In his right hand he held a reed, plucked from the banks of the River Tigris, which he was using to poke Abu Laith's newly acquired lion cub, who was asleep in his cage.

The man's wife and the rest of his children stood nearby, watching sullenly. Despite his efforts, the poking was having no measurable effect on the lion, who wasn't moving at all. All of this registered in Abu Laith's mind as he ran at full pelt through the zoo towards the man, who had not seen him coming.

It was around 7.30 p.m. in the zoo by the Tigris, and the dusk was settling pink over Mosul's Old City. Families were sitting outside the zoo cafe drinking cold Pepsi and glasses of tea. The bears were reclining in their cages as Abu Laith charged past.

'What are you doing?' shouted the self-appointed zookeeper, who rarely spoke at less than a bellow. 'Get out of the zoo.'

The man, who did not realize the danger he was in, barely glanced up. 'Why aren't they doing anything?' he asked, irate. 'We paid money to see them.'

Abu Laith came to a dead halt in front of the family. 'They're full,' he shouted. 'They've just eaten. When animals are full, they sleep.'

The man, who wasn't listening, kept poking at the lion cub. Next door, the lion's mother and father – known to the zoo's employees as Mother and Father – were also asleep.

'We paid money to see them move,' the man said, prodding the lion cub again.

'How would you like it if I poked your children with a stick?' Abu Laith spat, advancing on the family.

The man, who had finally got the message, backed away, his wide-eyed family backing with him. 'I'm not coming here again,' he said, snippily.

'Good,' called Abu Laith, as the visitors turned and scuttled off. 'And you had better not, because if you do I'll feed you to my lion.'

Grumbling to himself, Abu Laith turned his attentions to the cub. He was sound asleep, and looked not unlike a middle-sized ginger dog. None of the zoo workers, who were milling aimlessly around the park, had reacted to Abu Laith's outburst. They were used to it.

Everyone always said that Abu Laith himself looked like a lion, and it was true. He was five foot six with a rock-hard keg of a belly and an opaque halo of orange hair. His nose looked like it had been hewn from a boulder and sprinkled with freckles. He spoke in a roar.

That was why they called him Abu Laith, which – loosely translated – meant Father of Lions.

Since he could remember, Abu Laith had loved animals, and devoted himself to them at the near-absolute expense of humans. He had raised dogs, pigeons, rabbits, cats and beetles and held them in his hands when they died. For his third-eldest daughter's birthday, he had driven a herd of sheep into the family home. He had once given a baby monkey a shower in his garden.

He had one ultimate, lifelong ambition: to live on a farm with large predators roaming free around him. In Mosul, this was considered a suspect ambition. It had, possibly, something to do with the restrictions on animals in the Quran and the Hadith. In the holy texts, dogs were listed as *haram* – forbidden – along with pigs, donkeys, wolves, glow-worms (and all such bloodless animals), snakes and chameleons (animals that have blood, but whose blood does not flow).

Most people, even if they weren't religious, thought that dogs were dirty, and somehow unsavoury, in the way that people in Europe felt about rats: plague carriers and unclean beasts that defiled their surroundings. Though some families kept pets, it was considered disreputable to own a *lot* of animals. Among the people of the great city by the Tigris, animal lovers had a shady reputation as hustlers, fighters and panhandlers. Pigeon breeders, a fraternity to which Abu Laith also belonged, were especially dodgy.

Under the Iraqi legal system, pigeon owners were not considered trustworthy enough to testify in court. They had a reputation for always getting into fights and drinking too much whisky. Abu Laith fitted the stereotype all too well. He was a *shaqawa* – a kind of good-hearted neighbourhood thug. The sort of man you might call if you needed an extra pair of fists in a fight, or if someone was harassing your daughter, and needed to be scared off. He would never let anyone else pay for lunch, and always lent money to his relatives, grasping as he thought they were.

Since he was a young man, Abu Laith had made his living as a mechanic, fixing cars in the neighbourhood. At first he'd earned a few dinars here and there, but now he ran a big garage with several employees, where he charged hundreds of dollars to fix large American cars of the kind favoured by Mosul's elite. But this was nothing more than a distraction from his real love: large, dangerous animals.

In 2013 he had decided to take an enormous step, which he hoped would change his life for the better. He was going to build his own zoo: a wide, open space with a park for animals to roam, and offices and apartment blocks that looked over it. It didn't matter that the city was plagued with suicide attacks and kidnappings. In Abu Laith's mind, the development would be a lot like Dubai – slick skyscrapers and open lands on the bank of a mighty waterway, albeit the Tigris rather than the Persian Gulf.

As he gathered together his funds, wrenching money back from tight-fingered relatives and strong-arming investors, he had searched for a plot of land. He discovered that a large swathe of grassland on the eastern bank of the Tigris was for sale. There was already a zoo next door. To Abu Laith, the plan seemed fated to succeed, as long as the two businesses could combine into one large zoo. Once he had finished building the new development, he could buy the animals from the existing zoo, adding new ones if he needed to.

One bright morning, he went and spoke to the owner of the zoo, a rich man from Mosul known as Ibrahim, who lived in Erbil, a Kurdish city 50 miles to the east. Like most wealthy people from Mosul, Ibrahim hid his money, knowing it would be a magnet for kidnappers and spongers. When he travelled around Mosul, he went in a simple taxi. He wore poor-quality clothes, rather than fine suits. Abu Laith understood this, and understood how he could be of service.

'I know you can't be here to keep an eye on your business,' he told Ibrahim, when he went to see him. 'But I know animals, and I know Mosul. If we work together, I'll make sure your animals are looked after well. Then we'll expand it together, and we'll make some money.'

As it was, Abu Laith knew very well, the animals at Ibrahim's zoo were in a pitiful state. He had been to scout it out a few times, and had been appalled at what he saw. The bears – a Syrian brown bear called Lula and her mate, who wasn't called anything at all – were tetchy and worried by the fireworks that were set off to entertain visitors nearly every Friday evening near the zoo. The ponies were skinny, and the lions in their metal cages, about the size of a car, were bored and left roasting in the sun.

Abu Laith decided to step in and transform the lacklustre park into a proper zoo. With Ibrahim's blessing, he began to visit the animals after he finished at the repair shop. Abu Laith, despite never having been a zookeeper before, had spent his life preparing for the role. From hours of watching the National Geographic channel, a years-long obsession of his – it played uninterrupted in his Mosul home – and from owning dozens of pets, he had accrued zoological knowledge that he considered unparalleled. When he was unleashed on Ibrahim's zoo, it was as if a bomb had fallen from the sky. The zoo employees quickly learned to shuffle off when they saw the portly red-headed man stalking towards them. He would inevitably be getting ready to shout at them for not having cleaned the cages, or for feeding barley to the lions.

'They need meat,' he would spit. 'Fresh meat, only just dead.'

Abu Laith was in his element. Soon, he hoped, he would have raised enough funds to start building his own park on the plot of land he had bought next door to the zoo, which for the moment lay empty. When it was done, these animals would be able to run free, rather than being cooped up in those small, hot cages.

It would begin with the lion. By early 2014, Abu Laith had for six months been the proud owner of a lion cub. The little lion had tawny orange fur, and a notch on his upper lip where he had caught it on some chicken wire that Abu Laith had ill-advisedly used to protect his cage from stick-wielders and other disturbers of the peace.

The lion cub was his first acquisition for the new zoo – the first animal that would be truly his, and not Ibrahim's. He had first met the cub in Ahmed's house, which lay about half an hour east of the Old City. Ahmed worked at the zoo, and he infuriated Abu Laith, who disliked the way he always wore tracksuits and his disdain for the proper feeding habits of animals.

For some time, Abu Laith had expected that Ahmed might be hiding something from him regarding the pregnancy of a lioness who had been brought to the zoo two years before with her mate. Abu Laith suspected that when the lioness gave birth, Ahmed would try to steal her offspring and sell the cubs without the knowledge of the zoo's owner.

While he might often turn a blind eye to some stealing, Abu Laith was not going to be cheated out of a lion. As a self-styled manager of Ibrahim's zoo, he had decided early on that he had a claim on the lion cubs, and had arranged to buy as many of them as he could once the lioness gave birth.

Though he had never seen them in the wild, Abu Laith knew a lot about lions courtesy of National Geographic. He knew, for example, that lions sharpened their claws on stones, and that they liked to sleep after dinner.

All Ahmed knew about, he thought, was money. So when one day the pregnant lion started looking a bit skinnier again, with no sign of the cubs, Abu Laith suspected immediately that something was up. Biding his time, he waited on the street outside his house until Ahmed's eldest son walked past.

'Son,' Abu Laith called nonchalantly. 'Do you know where your father is keeping the lion cubs?'

'They're at home,' said the boy.

It wasn't long before Abu Laith was parking his large American car outside Ahmed's house, a small building with a garage. Inside the garage sat Ahmed, who was peering into a modest brick structure containing two very small lions, each no bigger than a loaf of bread.

Abu Laith was furious. 'Why did you separate them from their mother?' he shouted, as he stormed into the room. 'Now if she sees them, she'll smell human on their fur, and she'll eat them.'

Ahmed, lounging in his tracksuit, didn't seem to care. 'Which one do you want, then?' he asked, clearly exasperated that he'd been rumbled.

Abu Laith crouched down and cast a professional eye over the lions. Moving slowly, so he wouldn't scare them, he opened the door to their enclosure. Immediately, one of the cubs jumped out and on to a white plastic chair that stood in the middle of the garage.

'This one is mine,' he declared, beaming at the young lion, who looked back at him calmly. Within a matter of days, Abu Laith had installed the lion in a cage next door to his parents in Ibrahim's zoo.

After a period of consideration, he decided to name the cub after the lion in a cartoon about African animals that he had watched with his children, complete with mistranslated Arabic subtitles. He would call him Zombie.

Immediately, he set to work training the lion. He taught Zombie to sit quietly outside the cage when it was being cleaned. When he told him to go back into the cage, Zombie would obey. The cub knew not to bother the other animals in the zoo. Across the way from Zombie lived the two brown bears, Lula and her mate. The male bear was admirably strong, Abu Laith thought, and very protective of Lula. When the zookeepers had once tried to move him into a separate cage from his female companion, he had roared and fought so much they had given up.

Lula was a quiet soul who liked honey. When Abu Laith finished up at his mechanic's shop, he would come to the zoo with half a kilo of honey for Lula, who would eat it and lick it from her paws. She liked apples, but only if they hadn't touched the ground. She was a very clean bear.

The training continued apace, and within a few months Abu Laith knew, with the confidence of a man who had only met four lions in his life, that he would be able to tell Zombie apart from a thousand others of his species.

When night fell, and all the families and the small, annoying children were gone, Abu Laith would take a bottle of whisky to the zoo and sit down with Zombie for a yarn.

'If animals are really dirty,' he would sometimes ask, gazing out over the Tigris as the reeds rustled, 'why did God create them?'

The lion couldn't answer, but Abu Laith thought he knew what he was talking about.

A few months after Zombie came to the zoo, however, Abu Laith's dreams of building his own wildlife park on the Tigris were dashed by a suicide attack that killed one of his business partners in the zoo-building venture just as he was emerging from Abu Laith's front gate. He had been drinking with the man in his courtyard, and Abu Laith survived, but was accused by the police of having ordered his partner's murder.

Because the man had died in front of his house, Abu Laith felt compelled to pay compensation to his family to the tune of almost all his considerable fortune, amassed through years of saving up every dollar from fixing American cars. After four months in prison, when he was released after the police realized he wasn't a murderer, he came to the zoo to see Zombie, his dreams of re-creating Dubai on the Tigris in tatters.

He could feel the lion had missed him.

2

Hakam

BY THE TIME HE HAD TURNED TWENTY-FIVE, HAKAM Zarari was a seasoned weightlifter, a bird tamer and one of the Iraqi Ministry of Agriculture's most talented chemists. He could bench press over 120 kilos and had written his master's dissertation on the theoretical study of critical packing parameters of hydrotropes, using DFT theory and QSAR calculations. He had a pet bird called Susu who slept on his chest.

His family were all similarly overachieving. Hakam's parents, Said and Arwa, were lawyers and his sister, Hasna, was majoring in literature. She was twenty years old and studying English at Mosul University, an august institution of ochre stone on the eastern side of the city, where she read Shakespeare and Jane Austen. Hakam's dashing younger brother Hassan was away in the US studying for a masters in law at Penn State University. Their house was one of the mansions that lay on leafy roads not far from the eastern bank of the Tigris. Behind the thick peach-coloured walls that faced on to the street the garden was a verdant paradise: an orange grove banked with delicately tended flower beds, and beyond them a towering house with airy rooms.

Being part of Mosul's upper crust, however, did not insulate the family from the unstable and dangerous reality of their city. Mosul, a stronghold of Iraq's Sunni minority, had for years been under the strict control of the army, sent by the Shia-dominated government in Baghdad.

The soldiers had kept the city on lockdown in response to a wave of attacks from Sunni jihadis, part of a homegrown insurgency that swept the country after the American-led invasion of 2003. The jihadis attacked the American and British armies, as well as the local army they had created after dissolving Saddam Hussein's forces, with suicide and roadside bombs. Though Mosul wasn't as notorious as Fallujah – a city to its south known as the 'graveyard of the Americans' – it was plagued by violence. In 2004, al-Qaeda launched a takeover of the city, which was only put down after the intervention of thousands of Kurdish, American and Iraqi troops. For years afterwards, the jihadis retained enormous control over the city's western side.

But for many of Mosul's residents, the soldiers were invaders, rather than saviours – an occupying force. The suicide bombings continued, sometimes more, sometimes less, but leaving an ever-present fear of strangers and crowds. The soldiers seemed to delight in causing endless traffic jams and humiliating people at checkpoints. They set up roadblocks on a whim, conducted relentless stop-and-search operations and smashed the windows of parked cars that stood in their way. Those they arrested sometimes came back crippled from torture, or didn't come back at all.

One of Hakam's relatives was kidnapped from his workplace by a group of corrupt army officers in Mosul's main industrial district. At first, no one knew where he was. After a round of frantic calls it became clear that the kidnappers would give him back if the family paid a ransom. State-sanctioned kidnapping, the family reasoned, was infinitely better than being held on political grounds. They paid the ransom and he was returned, relatively unharmed. Little more was said about it. The man was lucky.

As the government tried to bring Mosul and its jumbled streets under control, the architecture of the city itself was altered, turned into a strange network of fortifications. The entrances and exits around the district where the Zarari family lived were closed off by roadblocks. The only way in and out was through a checkpoint at one end of the district, an area of about three or four blocks.

In theory, the aim was to stop the jihadis from launching multi-pronged suicide attacks. In practice, it inconvenienced the area's residents and provided more work for the soldiers – many of whom were already bored, angry and spoiling for a fight. Their friends had been killed by Sunni fanatics, and many thought the city's residents were little different.

The army checkpoints meant it took Hakam an hour and a half to get to his lab, a mile away. During the military lockdown, basic services were neglected: water was intermittent at best, the electricity flickered constantly and sometimes disappeared for hours, and in the slums around the Old City – which teemed with resentment towards the government – sewage ran in the streets. It was, many Moslawis thought, insultingly clear that the government did not care for the wellbeing of their ancient city.

While Sunnis had held most of the power under Saddam Hussein's supposedly secular government, the ruling class installed by the Americans was decidedly Shia-dominated.

Their new leaders were eager to exact revenge on the people they saw as their former oppressors. Despite the army's efforts, al-Qaeda cells regularly targeted the soldiers, and their American backers, with car bombs, suicide attacks and sniper bullets. More often than not, civilians were killed alongside them.

Every day was a risk. Life was normal one second, and the next, everything was dust and blood, eardrums broken, screams and chaos. If Hakam was at school, his parents would sometimes call to tell him not to come home because there had been an attack near the house. Before mobile phones, it would take hours before a relative returned home after crossing an area where a suicide blast had taken place. The family would wait, glued to the news, hoping that this time they would not be affected.

In 2005, when he was sixteen, Hakam was walking home one day with his friends from a study group. It was summer, and they had been taking private lessons ahead of their baccalaureate exam. They were meandering through the heat towards a checkpoint when someone started firing a gun just ahead of them. There was no shelter, no houses to take them in. They hit the ground as the world around them exploded. A huge thump shuddered the road in front of them. Maybe an armoured car had been blown up, Hakam had thought, as the debris fell around them, and he prayed he wouldn't die.

By now, Hakam knew the anatomy of an attack. Sometimes the militants would only sweep past a checkpoint, spraying it with bullets or detonating a suicide bomb before running away. If you were less lucky, you'd be caught up in an attack with a specific target: an assassination or an assault from different directions aimed at destroying a checkpoint. This was one of the targeted attacks.

For two or three minutes Hakam and his friends lay there, hands over their heads, waiting. The street was filled with smoke, the screams of the injured, the shouts of the soldiers. As it quietened, the boys stood up, terrified. The tables were turning, as they always did after an attack. Soon the soldiers would start blindly shooting at anyone dressed – as the militants were – in civilian clothes. Everyone was a target. The boys high-tailed it down the road.

When Hakam went back through the checkpoint the next day, there was no sign of the attack. There were families walking on the street, and people lining up to pass through, grumpy in the fume-soaked heat. They had all learned to live with it.

Across the city, opinion was divided: some saw the jihadis as gutsy liberators who would rid them of the army, others – like Hakam's family – saw them as troublemaking fundamentalists.

Waiting in the fifty degree heat one day at a checkpoint on his way to the gym, Hakam wondered what would happen if the soldiers left. He wheeled his pushbike over the uneven roadside towards the soldiers standing at the barricade. He would much rather have driven a motorbike, but they had been banned for years after becoming the transportation of choice for suicide bombers. Instead, he cycled along Mosul's traffic-choked streets on a green and blue pushbike, attracting strange looks and inhaling lungfuls of dust.

Without the checkpoints, it would have taken him five minutes to get to the gym. Now, because of the spaghetti-strand route he had to take, it took him a lot longer. Sometimes he would charm the soldiers, who would let him sail past.

He'd learned their names, so he could call out and say hi as he approached, buttering them up. But every two weeks the units changed, and a new – roundly suspicious – group of soldiers came on duty. They were nervous, jumpy and sometimes bent on revenge for friends who had been killed by the jihadis. Every Moslawi they saw was a potential terrorist.

This time, Hakam could tell, would be bad. As he rolled his bike towards the checkpoint, he saw unfamiliar soldiers at the barrier. He braced himself for an argument, and smiled pleasantly. Some of the soldiers were sitting on chairs, others standing up to check the cars. One walked up to him and made a cutting motion with his right hand over his left arm, the universal Iraqi sign for papers.

Hakam passed over his identity card. The honking of the cars was so loud it was giving him a headache. The soldier looked at the card for a moment, then stood back. Hakam held out his backpack, packed with well-worn gym gear.

The soldier rooted through the pants and socks. He pulled out a protein shake, unscrewed the top and looked inside at the milky swirl, checking for a bomb. 'What the hell is this?' he asked.

'It's a protein shake,' Hakam said. It was the same every time there were new soldiers. He adopted a tone of studied patience. 'I'm going to the gym. I live round the corner. I come here every day.'

The soldier looked down at the bike. 'Hands on the wall,' he said, pulling the bike away from Hakam. Around them, the cars blared long, insistent signals.

Hakam turned and raised his hands towards the wall. Men had been lost this way, taken from checkpoints and never seen again.

'What's your name?' the soldier asked.

'Hakam Zarari,' he said.

'Where are you going?'

'To the gym,' Hakam said, as calmly as possible. 'I come here every day. I live really close by.'

There was no reply. The soldiers had walked off – some to check the cars going past, some to shake down pedestrians, some to smoke and drink tea. Hakam waited, his hands on the wall. He didn't want to look round. His shoulders ached. He felt embarrassed, which was what they wanted. They were, he thought, ignoring him on purpose. Anger and shame coiled inside him as the sweat soaked through his t-shirt.

Soon, he began to wonder whether they'd just forgotten about him. The cars were still honking, and the air pressed even hotter. He chanced a look behind him. The soldiers were standing around the line of cars, looking through their windows and occasionally opening the boot and checking underneath the chassis with a mirror. No one was looking at him.

He turned back to face the wall. The soldiers were impossible to talk to. He would have to wait.

As the minutes passed, he sank into a heat-struck fog. This was worse than the usual treatment: being screamed at and called a son of a bitch by the soldiers.

'Hakam?'

Someone shouted his name from across the road. Keeping his hands on the wall, he turned around. His cousin Mustafa was standing opposite the stream of cars snaking in both directions, looking extremely confused.

Mustafa was a student about a year younger than him, pale-faced and cheerful. They'd planned to go to the gym together that afternoon. He seemed to ignore the group of armed men imprisoning his cousin. Cutting through the swathe of traffic, he ran up to Hakam.

'What is going on?' he asked, as much to his cousin as to the soldiers.

The men looked up and sauntered over. Mustafa handed over his ID card, and the young men stood together, staring at the soldiers.

'So you know this guy?' one of the soldiers asked, sounding extremely bored.

'Yes,' Mustafa pleaded. 'He's my cousin. He lives just down the road.'

The soldier mulled things over for a moment. 'Fine,' he said. 'You can go.'

A few minutes later, Hakam was back on his bike, shaking like a tuning fork as Mustafa followed him away from the checkpoint. He'd been there for almost half an hour.

'Screw this,' he thought, cycling towards the gym.

3

Abu Laith

The thing about lions, Abu Laith considered one afternoon as he went to visit Zombie at the zoo, was that it didn't take long at all for them to grow up. In his years of watching the National Geographic channel at home in his living room, he had observed the speed with which lions went from small, helpless creatures to large ones that could kill a human with a swipe of their paw. The key, he knew, was their mother's milk, which made them grow at an astonishing rate.

But since Ahmed had separated Zombie from his mother the cub couldn't live in a cage with her, and he couldn't feed on her milk. Abu Laith knew that if Mother had smelled a human touch on Zombie, she might kill him. Because of Ahmed's ignorance, he had been left to raise the lion with his own hands. But he didn't know what to feed him, other than cow's milk.

The result was that, despite being six months old, the lion cub – who lived in a small cage next door to his parents on a patch of cement about 20 feet by 30 feet – was still the size of a large puppy. Zombie wasn't growing, and Abu Laith didn't know how to fix it. This vexed the would-be zookeeper immensely. No matter how much he fed the lion cub, he didn't put on weight. Abu Laith bought Zombie's milk in big bottles from a shop near the illegal sheep market by the city walls, and fed it to him from a bottle with an extra-long teat – the kind, he knew, used to feed lambs that had been rejected by their mothers. Zombie usually gulped it down, spilling frothing white foam around him, and over Abu Laith's trousers. Yet despite this, he remained resolutely small, and rather timid, for his age.

Today he looked the same as ever. As Abu Laith walked through the zoo, sweating in the burning sun, he saw Zombie sitting in his cage, small and red as a fox. Abu Laith had just finished a full day at the mechanic's shop, and was covered in oil, keen for mischief. It was May 2014, not long after he'd been released from prison. The weeks inside had been boring, as well as an insult to his reputation. But now he was free, his name clear, and freedom suited Abu Laith well.

As he trundled around the city in his ostentatiously large truck – big enough for all his many children – he didn't think too much about the suicide bombs and the kidnappings that plagued Mosul, and that had taken a fair handful of his friends and relatives. Abu Laith wasn't prone to introspection, and he saw nothing to be gained from worrying all the time, as his wife, Lumia, did. Though he hated the trappings of organized religion with a deep passion, and thought mullahs were no more than bearded hypocrites, he knew his life was in God's hands. He would die when his death was written, and not before.

Zombie's life, however, was in Abu Laith's hands. Something would have to be done if the lion was going to grow up strong. Abu Laith had big plans for Zombie. By the time he was older, he hoped, he would have built him an enclosure so big he could hunt – running down sheep and goats in a safari-style environment that Abu Laith did not think would be hard to arrange in Mosul.

'You're too young to eat meat,' Abu Laith told the lion, surveying him critically. 'And you don't like honey.'

Attempts to get Zombie to eat a more varied diet had gone badly, with the lion turning his nose up at everything Abu Laith had offered him. Apples and bread were sniffed at and ignored, as were thin strips of goat flayed lovingly from a fresh carcass brought from the slaughterhouse near the zoo on the eastern side of the city.

After a long period of contemplation – and many meditative hours in front of the National Geographic channel – Abu Laith had decided that if cow's milk was not enough, then Zombie needed donkey's milk. He had heard somewhere that the queen of Tadmur, the ancient city in Syria guarded by statues of two winged bulls, had washed her face with it. She was famously powerful, and it seemed reasonable to think that it might make Zombie strong.

He had asked the farmers at the market, and the raggedy boys who rode donkeys through the streets of the Old City. But after days of searching, and enquiries posted in every dairy inside the city walls, he gave up, unable to find a regular supplier. There just weren't enough milk-producing donkeys when he needed them.

That was weeks ago, and he hadn't had a better idea since. Zombie was still living on two very large bottles of cow's milk a day, and his belly was covered in shrunken ochre fur, ribs faintly protruding. Now, he was hungry again. Downcast,

Abu Laith opened the door of the cage and walked in to give the lion its supper.

'Sit,' he said, and Zombie sat. The lion was always very obedient, Abu Laith recalled later, except on the occasions when he didn't feel like listening. Zombie's training regime, which Abu Laith had concocted over hours in the mechanic's shop while tightening rivets or changing oil, was taking hold, even if the food wasn't making him fat. Inspired by a video he had seen about tame lions, Abu Laith had semi-successfully taught Zombie to sit outside the cage when he cleaned it, and to come to him when called. The fact that the cub didn't always obey these commands was simply proof to Abu Laith that he was an independent-minded lion, rather than one who did not follow orders.

It was late afternoon by now, and Abu Laith was cleaning out the cage when he saw the buffalo herders cross the bridge over the river, and had an idea. Discarding his broom, he shut the cage door and raced through the zoo and along the Tigris towards them. They were a group of three or four men, some dressed in long thobe gowns, holding long, thin sticks to guide the lumbering animals on up the road, where the slow-moving traffic jam honked itself forward in a fug of exhaust fumes.

'Do you sell milk?' Abu Laith shouted at the men.

'We do,' replied one of the herders. 'You're welcome to it.'

Amid the cars, with the mud-banked Tigris stretching out below them, the herders poured out a stream of foaming liquid – shining like oil – into repurposed Pepsi bottles. Pleased at his own ingenuity, Abu Laith stacked them in his arms and started back towards the zoo, as the buffalos plodded away down the road.

Zombie hadn't moved far. He was still sat in the cage, looking at Abu Laith. The zookeeper poured a large slug of buffalo milk into Zombie's bottle. Holding the lion still, Abu Laith hallooed as he gulped the milk down. He later explained his thought process: if this milk could make small buffalo calves turn into the huge beasts that lumbered down the highway, it should have a similar effect on Zombie.

The experiment was an unqualified success. With at least 2 litres of buffalo milk a day, Zombie began to pile on weight at a remarkable speed. As the summer grew hotter, his stomach rounded, his neck filled out and he seemed, to Abu Laith, content with his lot. Next door, his parents alternately paced around the cage or slept. To the intense frustration of the visitors to the zoo, they almost never roared.

Across from them, Lula the bear and her mate lazed in the sun, or licked honey from their paws. The Shetland pony trotted around its enclosure or, occasionally, went berserk when Abu Laith tried to put a halter on it. Ahmed skulked around at the entrance, taking money for tickets and spending not much time at all on the animals. From Ibrahim, the animals' owner – who lived in Erbil – they heard very little.

All the while, in his self-appointed role as zookeeper, Abu Laith kept a firm hand on the comings and goings at the zoo by the Tigris – seeing that the pony was given hay, and that the monkeys had branches to swing on. Soon, Abu Laith thought contentedly, as the summer heat rose, everything would calm down, and he could really get working on his own zoo.

4

Hakam

At 12:30 p.m. on 5 June 2014, a ringing bell made Hakam look up from his computer. His supervisor was shouting for everyone in the lab to go home early, and around him people were picking up their bags and sloping off towards the door. He glanced out the window, and saw the streets thronged with people and a massive traffic jam.

Hakam made his way outside with his colleagues, crowding each other on the stairs. He didn't think too much of it. Government employees were usually sent home when there was a security threat, as there often was. They filed out into the car park, each clutching their briefcase. His phone rang. It was his father.

'Did they tell you what's happening?' he asked.

Said Zarari was a lawyer, one of Mosul's best. He worked at the courthouse about five minutes away from Hakam's office, a large white building filled with dusty files. He sounded extremely terse, which was not unusual.

'No idea,' said Hakam, as his colleagues piled into their cars and joined the slow traffic jam. 'What shall we do?'

'I'm leaving work now,' said his father. 'I'm trying to see which road has the least traffic. Come over to where I am.'

As his colleagues left, Hakam looked out at the teeming streets, which were filling up with office workers who had been sent home. No one seemed to be panicking, though the traffic jams and the army's attempt to direct them were goading drivers beyond endurance. Shopkeepers were waiting at their doors for their last customers to leave, before pulling down the shutters with a clattering bang. Mannequins were covered up, and the shops padlocked.

There were a lot of soldiers around, shouting incoherently at drivers to stop or go – often both at the same time – and shooting in the air. As he waited, Hakam recalled later, he had wondered if they were preparing an operation against the jihadis. For the last few days, Sunni fundamentalists had been battling the armed forces in the city's poor, conservative districts on the western bank of the river, bolstered by militants who had taken control of much of Anbar province to the south earlier that year.

It said on the news that the jihadis were overwhelming the soldiers in the west of Mosul, though the army was insisting it had the situation under control. Few Iraqi soldiers, who mostly came from the Shia areas to the south, wanted to die defending a Sunni city. They had lost friends to suicide bombers, and saw the locals as dangerous fanatics.

Little quarter was given to the citizens of Mosul, who were harassed at checkpoints by the army and killed by the random bombs of the self-proclaimed holy warriors – stuck between the lines of a war where both sides were claiming to be their protectors.

The two sides of Mosul were separated by the churning River

Tigris, about 200 yards wide as it cut through the city centre. The east was, in general, slightly better off, and better staffed with army units. In the rickety streets of the west side, there was some local support for the jihadis. But most people just wanted to be left in peace. Hakam loved the Old City on the west side – the bustle of traders and the smell of roasting coffee from gold-painted stalls that dotted the streets – and he was proud that centuries of Moslawi history were layered in the stone alleyways along the western bank; invaders from Hulagu Khan to the Ottomans and the British had fought and died there.

At times, over the years, the jihadis had turned the Old City into a viper's nest, hiding in alleyways and rooftops to launch surprise attacks on the army. At night, some neighbourhoods in the west were held by jihadis. It was all the more dangerous because of how embedded they were in local society. Three brothers, living in the same house, could be respectively a soldier, a policeman and a militant. All it would take was for the jihadi to ask his brothers not to go to work the next day, and a road would be left wide open for him when the other two warned their colleagues to stay away. With the geography of the Old City on their side, a few hundred guerrilla fighters could easily resist and immobilize a vastly more numerous and better-equipped enemy.

The east was different: it had broader streets and the buildings were squat and relatively modern. Though there were many there who sympathized with the jihadis, it was generally richer and less insular than the west – more people who worked for the Baghdad government lived there, and fewer lived in poverty and humiliation.

Connecting the two sides were five bridges. One, known as the Old Bridge, was an iron masterpiece built by the British during their occupation of the city after the First World War. On any given day, the bridges were gridlocked with traffic going back and forth from the markets on either side, or ferrying Moslawis to school and work. There was, Hakam thought, no chance the jihadis would be able to cross the bridges to the east.

His father called him. 'You'll have to walk over to me,' he said. 'The traffic is too bad.'

Winding through the cars, which were all but parked in the road, drivers leaning on their horns, Hakam made his way to his father. Said was dressed, as ever, in a pristine suit and an astrakhan hat that gave him an air of deep sobriety. 'Come on,' he said. 'The traffic is terrible.'

It was worse than they had expected. As they crawled along the road eastwards, Hakam saw soldiers gathering everywhere at checkpoints – a scrawl of blast barriers, sandbags and Shia flags emblazoned with the face of Imam Hussein – or rushing in armoured cars towards the west.

'I don't like this,' said Said, and Hakam agreed.

It took them four hours to get back to the house, which was only a few miles from the courthouse, through traffic that seemed to get more desperate at every turn. No one wanted to be stuck on the street, sitting ducks for the jihadis.

They parked the car outside the house. A metal door led through the 10-foot high outer walls. Inside was the courtyard, which at this time of year was planted with red and yellow flowering shrubs, shaded under the orange trees that grew heavy fruit in the winter. It was like many houses in Mosul: from the outside, a dusty wall, but inside, a lush garden and high-ceilinged family house full of life.

Arwa and Hasna, Hakam's mother and younger sister, were

already in the living room, listening to the news in silence. Next to them was the family's collection of books, an entire wall of leather-bound volumes and paperbacks stacked tight on dark wooden shelves.

'You're home,' said Arwa, deeply relieved. Like her husband, she was a lawyer, though she hadn't practised since she married.

Hasna was hunched on the sofa in her usual trousers and long shirt, glued to the TV. The anchor was reporting that an operation was going on in western Mosul to neutralize terrorists.

'They'll stop them,' Said said. But Hakam could see that his father, who terrified criminals in court, was sober and doubtful.

Hasna, who had been at university getting ready for her English literature exams, asked 'What are we going to do if they come?'

'We'll stay,' said her father. 'We've got everything we need here. They wouldn't care about people like us. They only want to hurt the army.'

In the tense atmosphere, no one argued.

The family sat down and waited. None of them were sure who these jihadis were. The people the army were fighting in the west could be al-Qaeda remnants, or the groups who had taken Fallujah and Ramadi. They were backed by the jihadis who had taken Raqqa and parts of Aleppo, defeating the more secular Syrian rebels in the terrible civil war that had been going on since 2011.

All evening, the TV ran updates on the fighting in Mosul, interspersed with sports news and the weather in New York and Washington. The Iraqi channels knew no more than their foreign counterparts. The family's phones rang relentlessly with news and questions from relatives who lived outside the city, or those who were waiting nearby, frantic to know what was going on.

Hassan, Hakam and Hasna's brother, stayed dialled in on video call from the US, his terrified face pale on the screen that stood propped up on the coffee table. Hassan had left for the US just a few weeks before. Now he sat in his dorm room, tuned in to every channel he could. But for all that the family talked, they still heard only rumours and the occasional distant explosion.

At about 2 a.m. they were still sequestered in the living room – splayed across the sofas, half-dozing in front of the TV. The honking outside, which had been going on all evening, was growing louder.

'Can you hear them?' Hakam asked, walking towards the door.

'Don't go outside,' said his mother, sharply. 'We don't know what's happening.'

Hakam considered this. 'I'm going to find out,' he said, and ran up the stairs. On the landing near Hasna's room, there was a window that looked out on the street. Crouching, he stared through it, trying not to create a silhouette. He could just see the big intersection that lay behind the house, with its watchtower on the far side always manned by at least two soldiers.

Usually, the road was empty at this time of night. But now it was rammed as if it was rush hour: cars beeping and drivers crawling forward, tail to tail, the line of vehicles almost stationary. With a rush, Hakam realized what was happening. This was the road east. They were escaping Mosul – probably heading for the safer Kurdish cities, if they had the connections to get through the checkpoints.

To the right and left of their house, Hakam could see his neighbours packing their possessions into their cars, the lights on and faint shouts audible over the traffic.

But Hakam's father had been clear. They were not leaving.

By the next evening, the family had all but stopped talking. They had gone through every possible scenario so many times that each option seemed about as unlikely. None of them had gone outside. The TV hadn't reported anything new, but the bangs in the distance had grown louder.

Hakam went back to his vantage point at the window. The traffic jam had broken up, though there were still plenty of cars on the road. Among them there were – Hakam saw – a surprising number of the army's beige armoured vehicles driving out of the city. None were heading in, though he could still hear shots cracking in the distance. He looked at the watchtower on the intersection. For the first time, ever, it was empty.

Hakam sat back on the floor. His brain felt numb, he later said. The soldiers were abandoning Mosul – running away from the jihadis. But it didn't make any sense. The streets of Mosul were usually choked with soldiers. They had huge bases with 15-foot blast walls. They had tanks, machine guns, helicopters.

But though the army had enormous garrisons, built and armed by the Americans, and thousands of men, they were weak and demotivated. The jihadis might only be a gaggle of extremists, but they were full of religious bloodlust, and hundreds of them in their dusty pickups came in from the desert, with their Kalashnikovs and rocket-propelled grenades.

They had taken Anbar to the south like this, and Raqqa in Syria – rising together to displace Iraqi government or Syrian rebel forces in a way that no one had ever really thought possible.

Whichever way Hakam twisted it, the situation looked the same. Not much more than a thousand resistance fighters armed with rifles, suicide cars and machine guns mounted on pick-up trucks had launched an assault that splintered a thousands-strong fighting force trained by the US army at the cost of billions of dollars. Across Mosul, the Iraqi police and army had abandoned their posts.

He walked downstairs, to where his family were still sitting. 'The army have retreated,' he said in disbelief. 'I saw their armoured cars leaving. They've abandoned the watchtower.'

The jihadis had won, and they had barely needed to fight.

By the morning of 10 June, less than a week after the militants had begun the fight for Mosul, it was quiet. Hakam walked outside. The streets were empty, and everything looked the same. But the Islamic State had taken power in Mosul, Iraq's second city and home to over half a million people.

5

Abu Laith

About three days before the jihadis had come, Abu Laith had gone drinking with two of his friends in Bashiqa, one of the Christian villages that lay on the Nineveh plains outside Mosul. They had spent the day sitting in their usual spot under a sumac tree drinking Black Jack, the horrifyingly acidic local whisky, and cloudy, cold glasses of aniseed arak. When Abu Laith had tried to drive home that evening, however, the road had been closed. The soldiers had said there was a curfew, and that he couldn't go into the city.

Abu Laith had grumbled, but there was nothing he could do. It was unusual for the army to block the roads, but not particularly remarkable. Curfews happened whenever there had been a big terror attack, or when there was a threat of one. The workers at the zoo would feed Zombie. His older children, who were in the house in Mosul, could look after themselves and the younger ones for the night. Still reeling from the drink, he had gone to Gogjali, a suburb of Mosul outside the curfew zone where his wife Lumia happened to be staying at her parents' house with their three youngest children

– Nour, four, Ashraf (known as Geggo), three, and Mo'men, a baby whose hair was as bright a shade of sienna as his father's.

It took a few days for the road into Mosul to open again, by which time Abu Laith was so bored that the news came as an overwhelming relief. Lumia's family all spoke to each other at the same remarkable volume at the same time, no one listening for a moment to what the other person was saying. There had been nothing to drink, and no lions to talk to. Abu Laith was very worried about Zombie. Occasionally, they had gone out to check if the road was open, or to call Luay, Abu Laith's second-eldest son, who was in the house in Mosul with three of his siblings. For the last few days, they had been quite as bored as Abu Laith. There were rumours, and the sound of distant fighting, but that was all normal for Mosul, and no one thought much of it. They stayed at home, like they always did when it was like this, and waited for the latest crisis to pass.

By the time Abu Laith and the others had piled into the car on the third day it was well into the evening, and there was something very strange going on. The road into the city was completely open, while the road coming out was thick with cars.

'They're all going the other way,' Lumia said, staring out of the window. 'They're all leaving.'

'It'll be fine,' said Abu Laith, who was – nonetheless – a little concerned. 'Don't worry.'

'I'm not worried,' she snapped.

Being married to Lumia, Abu Laith sometimes felt, was like being married to a box of firecrackers, or a young bear. She was very exciting to be around, but could explode and kill you at any moment.

Lumia was a thirty-five-year-old firework of a woman who talked like a machine gun and laughed at almost everything. She had coal-black hair that she occasionally half-covered with a scarf, and told dirty jokes that made her children cry with laughter and embarrassment. When telling stories, as she often did, she was prone to re-enact them to the point that a casual visitor to her living room might find her crouched on the floor, pretending to be a rheumatic old man from Baghdad.

She and Abu Laith had met four years earlier, both widowed and burdened with grief and children (Lumia had three young kids, Abu Laith eight older ones). To their mutual surprise, they had liked each other with a messy, hilarious sloppiness that suited them both. Working quickly, they had knocked out another three offspring, Nour, Geggo and Mo'men, close as a relay race. Their numbers were supplemented by a remarkable array of animals; some recognizably pets, and some – like the beetles the children dug from the garden, or the sheep they had found one day and brought home – mere hangers-on, who were welcomed just the same.

Their crowded house lay across from the old amusement park built by the Americans in the east of Mosul. It was there, as his parents and siblings were returning, that Abu Laith's twenty-year-old son Luay sat at home staring at the TV with a creeping sense that something might be seriously wrong. He did not usually watch the news. Luay was in the middle of his exams. He was technically studying geography at Mosul University – a sprawling complex of institutional buildings further in towards the city centre – but was in practice spending most of his time playing a game called Clash on his phone or smoking narghile – a water pipe – on the front step of the shop next door. He was short, like his father, but powerfully built, and he had the same lion's head and easy manner.

But this time, he really was quite worried.

'A group calling itself the Islamic State of Iraq and Syria has temporarily infiltrated western Mosul,' said a presenter on one of the Arabic channels. 'The army is returning fire, and an operation is ongoing.'

Luay couldn't hear any fighting. The summer air was hot and still. He had heard of the Islamic State – one of the many Islamist groups that were gaining power in Syria, hijacking the more moderate opposition as well as Islamist groups and taking control of towns like the provincial backwater of Raqqa. In Iraq, they had gained ground in Anbar. But that wouldn't happen in Mosul, where the highly-trained army held the streets in a choke hold, and had spent years fighting fanatical groups just like this one.

Still, something wasn't right.

About 10 miles away, Lumia and Abu Laith drove towards the outskirts of the city and saw an empty metal shelter with a few blast barriers in front. 'The checkpoint isn't there,' shouted Lumia, and this really was confirmation that something terrible had happened. They drove further into the city. The checkpoint on the main road not far from their house – where they were usually stopped by the soldiers – was empty, too. They saw guns abandoned on the floor inside the hut.

'The army has gone,' Lumia said. 'They've run away.'

Abu Laith was not so sure. 'It could just be this checkpoint,' he said. 'The army are probably in the west.'

But as they drove towards their home, they saw only a stream of people going the other way. There were no soldiers, and the road into Mosul was empty.

It was 11 p.m. by the time they pulled up outside the house, and the night was black, the side street where they lived quiet. Abu Laith and Lumia ran in the front door, carrying the toddlers.

Inside, the older children were in the latter stages of panic. There was a pile of bread on a cloth in the living room, and Lubna – Abu Laith's third-eldest daughter, recently divorced and very harried – was in the kitchen. Luay was rushing around the house, trying to find clothes and blankets for the kids.

'We have to leave,' Lubna shouted, as she packed still more bread. 'Everyone else has gone. I think we're the last ones in the street.'

'The policeman down the road left hours ago,' said Luay, whose long hair always hung in his eyes.

Abu Laith took stock, surveying the chaotic scene in front of him. 'Right,' he said. 'We're going.'

The children were crying, Lubna was crying, Lumia was crying and Luay was still wearing his slippers by the time they had all got in the car – Abu Laith's enormous black and silver Chevrolet – along with the bread, some blankets and a few clothes. Lumia hadn't stopped talking, and Nour, who was as loud as her mother, had taken to shouting and gesturing in a manner reminiscent of Lumia herself.

'Zombie is in danger,' said Abu Laith over the din, grief-stricken. 'Who is going to look after him? He needs to eat. I need to go to the zoo.'

'We need to go,' shouted Lumia, and hustled her husband into the car. They had decided to go to Abu Laith's friend's house in Bashiqa, where just a few days before he had been eating mezze and drinking whisky. It was far enough outside the city that it should be safe from whatever this invasion was, Abu Laith had reasoned. The Kurdish Peshmerga fighters, who patrolled the edge of their territory, not far from Bashiqa, would protect the Christian towns if nothing else.

Packed tight into the car, they turned out of the quiet side street on to the main road, joining the slow-crawling traffic jam. Though the road wasn't wide, it had morphed into a multi-lane carriageway by the time they reached the outskirts of the city as thousands of people all tried to overtake each other at the same time.

A few miles outside Mosul, the traffic gridlocked and they stopped. All the drivers around them were leaning on their horns and jabbering. Those who could had alighted from their cars and were craning to see what was going on ahead of them. The night air was hot in the car, and diesel fumes went to their heads. In the back, the children were crying.

Shouting for everyone to be quiet, Lumia tried her door. It wouldn't open. The car next to them was too close. She reached over to the other rear passenger door. It opened for a moment, then slammed shut and wouldn't open again.

'If there's a fire we're all going to die here,' she shouted over the children's cries. 'We have to get out. We have to get out.'

Gunfire rattled somewhere in the distance. Around them, everyone was running, hauling themselves over the gridlocked cars with their families, bags abandoned. Abu Laith, who was already halfway out the car on his side, tried to see what was happening, but the night was dark and mad and chaotic.

Lumia, with considerable agility, shimmied over the front seats and out through the front passenger door. With Luay, she formed a human chain passing the children from the back seats into the front and out of the car, where they were corralled by Abu Laith.

Huddled together in a pack, everyone holding a hot, tired child or hanging on to someone else's arm, they made their away across the stagnant river of cars. Around them, people were shoving and shouting, looking for lost family members and trying to make it to a large warehouse that stood near the side of the road. Without really knowing why, Abu Laith and his family were drawn in by the wave.

The gunfire was louder than before, and seemed closer. Hundreds of people were in the warehouse, and all the children could see were the legs and the bags of the people around them. Those who weren't crying were shouting, fruitlessly, to find out what was going on.

At the front of the warehouse there were large CCTV screens that showed the river of cars and the people running from them. 'That's my father,' one man shouted.

More gunfire came, more shouts, until, for no particular reason at all, the family felt movement all around them and they were pushed out, streaming back to their cars. Luay had lost his slippers, and was running barefoot over the broken ground.

Lumia had kept up, dragging her skirts around her, until she looked down and, in a moment, was alone among strangers.

'Abu Laith!' she shouted. 'Luay! Lubna! Abu Laith!'

They were gone. Each person looked a little like them – children with blankets or short, stocky men, but they were all strangers. A burst of gunfire sounded right behind her and she ran with everyone else towards the cars.

'Abu Laith,' she shouted, and she ran into a man in military uniform. He was a Kurdish Peshmerga soldier, and he told her to calm down.

'Calm down?' Lumia said. 'They're shooting.'

And they were. Someone was shooting behind her, very close, and the people around her were sprinting flat out, or throwing themselves on the floor. Lumia was running, and then there was someone falling in front of her, and she stubbed her foot on him and stopped herself from tumbling over. He was a soldier, like the Peshmerga fighter she had seen before, and he was covered in blood.

Lumia screamed, and kept running towards the cars. Some of the doors were open, and people were pulling things out from inside them – looting bags and radios. The shooting had paused, and she stumbled in among the cars. They were moving now, and she dodged between them, screaming the names of her family until she saw Abu Laith just a few metres away, shouting for her.

They arrived at Abu Laith's friend's house an hour or so later. His name was Sheikh Hassan Ali Beg, and he was a former advisor to the then prime minister Nouri al-Maliki – like him, a Shia Muslim. The lights at the farm were on, and the family was awake and frantic. On the living-room walls hung pictures of Imam Ali, the Prophet Muhammad's son-in-law, who is seen by Shia Muslims as his rightful successor. The sheikh's family were members of the Shabak – an ancient minority in northern Iraq distinct from Kurds and Arabs and persecuted by both at various times.

'We're leaving for Duhok,' said the sheikh's wife, as she scrambled about the house carrying bags and shooing children. That city, 35 miles to the north, was well within Kurdish territory and ringed by mountains, which would slow the militants' advance. Whoever was coming this way, they were certainly Sunni insurgents, who would slaughter any Shia they found as apostates and heretics.

The sheikh's wife looked terrified. 'You're Sunni,' she said. 'You can stay. But they'll behead us.'

'We should go with them,' said Lubna, when the family had gathered in the sheikh's living room. 'We should run. We shouldn't risk it. We have bread, and we can manage.'

Lumia interjected. 'And what do we do when the bread is finished?' she asked. 'What are we going to eat then? We haven't even got any money.'

'What do you mean?' said Abu Laith sharply. 'Did you not bring the money?'

Lumia's white skin blushed. She was in charge of the household finances, and had a habit of hiding bags of cash around the house so the others wouldn't spend them. But in the rush to leave, she had forgotten to remove the money.

'I was busy,' she shouted. 'I was scared.'

Abu Laith was nonplussed. 'Well, we had to go back anyway,' he said. The decision had been made. If they didn't go back, they would lose their house and their money to looters. They were Sunnis, after all, and he had never worked for the Baghdad government. They could hold out for a few weeks until the army came back and destroyed the jihadis.

It was a long night. No one slept much, as the sound of the sheikh's family packing and the flow of cars on the road kept jerking the family awake. When light broke, they all had breakfast together. Lumia called her first husband's family, who lived in west Mosul, which had been taken by the jihadis a few days before.

'It's fine,' they had said. 'There's nothing different. The only thing is that the army has gone.'

While the day was still cool, they gathered their things and prepared to leave. Lumia left their bread behind for the sheikh's family, who she thought would need it. They were rich people, who were leaving everything they had, and she felt bad for them.

'You'll be safe,' the sheikh said, as Abu Laith's children piled into the car. He was smiling, but his wife was quiet and scared.

Abu Laith settled himself into the driving seat, and the children squawked in the back. The sun hadn't yet started to burn, and the river of cars from the night before had disappeared. In their place was a steady line of vehicles driving out of Mosul at an almost leisurely pace.

'They're gone,' said Lumia, looking out of the window as they passed yet another abandoned army checkpoint. 'They've sold Mosul.'

A grey haze hung over the city as they drew closer. Smoke was rising into the air from fires that seemed to obscure the sky, even through the hot summer sun.

They first saw them in the Saddam neighbourhood, on the outskirts of the east. 'Look,' someone in the car shouted. 'Over there.' In the distance they could just make out two figures holding guns, standing at the side of the street. They did not look like soldiers. After an instant, they were gone, and everyone in the car was talking at once.

'Calm down,' shouted Abu Laith. 'They'll be gone soon. Calm down.'

But no one really did, and by the time they arrived on their street – driving past the abandoned checkpoint into their neighbourhood – the anxiety was thick and palpable.

'Assalamu aleikum,' cried Abu Laith. Their neighbours, who had left the night before, had also come back, and were unpacking their car. They called greetings back. The mood lightened as Abu Laith opened the gate to the house – which was still locked, as he had left it – and they piled inside.

While the children milled around, Lumia was immovable. 'I'm not leaving the house until this is over,' she said. 'And neither are the children. That is it.'

Abu Laith wasn't really listening. More than anything, he was worried about Zombie. He didn't know if the workers at the zoo had fed him, or if the buffalo herders had fled, too. But the zoo was far away on the banks of the Tigris, about half an hour's drive from his house, and he couldn't go there now. No one knew whether there was fighting in city centre, or whether the jihadis might kill civilians. He had to stay at home to look after Lumia and the children until things became clearer.

For now, he needed to know more. Leaving the children squabbling, he went up the stairs, all four floors of them, where dust rose in the air from the breeze block floor. He opened the door that led on to the roof and walked out on to the flat concrete space that looked onto the old amusement park in front of the house.

It was hot now, and the sun strong. The drift of toxic smoke turned the air into cloying, acrid soup. Something brushed against him. It was a piece of black rubber. Charred rubber was falling everywhere, like rain, and Abu Laith knew where it came from.

He walked up to the edge of the roof, keeping an eye out for the jihadis, and saw the burning tyres lining the road a couple of streets over. Not far from them stood two men next to a pick-up truck. Abu Laith snapped down into a crouch behind the parapet of the roof, hoping they hadn't seen him. The men were dressed in long brown shirts and trousers known as Kandahari outfits after the Afghans who usually wore them. They had guns hanging from their shoulders. A black flag, he recalled later, flew from the back of their pick-up truck. They were the Islamic State.

6

Imad

1970

In the big house in eastern Mosul near the Tigris, where the rising mist cools the streets, Imad – the boy who would one day be called Abu Laith – sat on the floor. He was ten years old, about 4 feet tall and already barrel-chested. In his arms were Bobby, a mangy hound about the size of a fridge, and a dog called Jonny, who was even bigger. They had met just minutes earlier, in a pile of rubbish where they were both scavenging for food. Imad thought they were wolves, and he was very pleased that he had found them. After a moment of consideration, he had named them after two characters from the textbook in his English class.

Imad's father was shouting, as he had been for some time. His mother was crying, as she had been for even longer. 'These are filthy animals,' his father snapped. 'They have diseases.'

Imad, sprawled comfortably among the dogs, was implacable. 'I'll take them to the vet if they get sick,' he said. 'And I'll take them for walks every day.'

'You'll do nothing,' he father screamed. 'You'll throw them out where you found them.'

Bobby and Jonny didn't seem too bothered by the tirade. Neither was Imad. For all that he didn't like being beaten, Imad wasn't afraid of his father. The shouting and the threats were repeated every time he brought back a new animal to the house. It had happened with the pigeons, which slept inside two ghee boxes he had nailed under the eaves of the roof.

When he had bought them from the garage by Syria Road, he had picked them for their beauty: Sabuni pigeons – bone-white with feather crowns over their feet. Then came the doves, with their green-glinting wingtips amid the grey plumage. They neither carried messages nor fought, as other birds did, but he liked to hold them in his hands and release them flapping into the sky.

His father had also shouted about his sister's cat, which waged a war of attrition with the other animals for space as numbers grew, and his brother Ziad's donkey, which he rode backwards, until he fell off and broke his arm. It had gone a bit better with the lambs; Imad had bought these with his own money and herded them around Mosul, beaming as they bucked around the streets, annoying the neighbours and bringing shame on his family, according to his mother. When they had grown older, they had been rounded up and slaughtered. Imad had sat in the garden and cried until spring came again and the new lambs were born.

Imad loved animals of all kinds, though the more dangerous the better. He knew how to trap spiders and how to train a dog to fetch. He could lure any cat, dog or rabbit just to sniff his empty hands.

Within every living being, he knew instinctively, there was

a personality, a life with needs and likes and things they hated. Except bugs. Bugs he wasn't so interested in. They worked best to entertain other animals – moths to be the playthings of cats near a bright lamp, or a cricket to be given to a hungry bird. He liked birds. Their nests were their houses, and he would beat up any of the local children who tried to steal the eggs.

The family home in the Zuhour district was his fiefdom. It sprawled over acres of garden and forest on the outskirts of Mosul, where the wide, quiet streets seemed far from the dust and clamour of the city. He burrowed and played in the grounds, where he waged a long-running feud with Nouri, the gardener, a stout, kindly Kurd from Akre who had lived an easy life until he met Imad.

Nouri took great care of the gardens, and considered it a personal insult when Imad and his gang destroyed his work. With great effort, he had forced Imad's chickens into a purpose-built coop with a canopy of pomegranate branches. But Imad, who wanted them to run free, would open the door as soon as Nouri's back was turned, sending them to peck across his perfect lawns.

When Imad's sheep ate the flowers he had spent months cultivating, Nouri had finally broken. Furious, he had climbed the stairs to Umm Sabah, Imad's grandmother, who had ruled the house and everything in it since her husband died. She had a wide face and a horrifying temper.

It was a lost cause. Umm Sabah had a soft spot for Imad, whose chickens gave her eggs in the morning and whose antics were generally diverting. 'Let him be,' she had said, and Nouri had stomped back down the stairs, even angrier than before.

Imad thought he could house-train anything. Wolves would be no different. These ones he had just acquired didn't live in the snow, like the animals he had seen in books, but they looked just like them. He was not going to pass up this opportunity. 'You have two choices,' he said to his father, after he had shouted himself into silence. 'Either Bobby and Jonny stay, or I'm never going to school again.'

His father looked like he was about to hit him. Then he turned and stormed out of the room. Imad was very pleased, he recalled much later, and grinned as the great dogs panted.

'You live here now,' he said to them.

7

Abu Laith

Lumia's pledge that she would not leave the house lasted about two days. Every morning, she'd waited at home with the children and the pigeons for news of the outside world from Abu Laith, who had been out and about in the neighbourhood. Without the usual flow of neighbourhood gossip she was bored, and getting progressively snippier at husband, children and animals alike. Bustling around the kitchen, banging her pans, she cursed at the animals swirling around her legs and the sullen teenagers glued to their phones. It had been quiet since they got back. The only time they'd glimpsed the jihadis – armed, bearded men in long Kandahari shirts – was when they came to a house a few doors down (the inhabitants had fled) and took their car, which belonged to the government.

To make things worse, it was Ramadan, and everyone spent the day hungry and furious in the heat. Usually, the fasting month was a holiday, when special series ran on the TV, often family dramas thick with blood and betrayal. The devout prayed, while the others used the opportunity to sleep all day until it was time to break the fast in the evening. But when

Ramadan fell in the summer, the long days drove everyone half-mad. Though their fast began each day with the dawn prayers, it didn't end until almost 9 p.m., as the sun seemed to hang static in the sky. For the truly religious people, it was an important test of their faith. For the less spiritually minded, it was incredibly annoying.

'That's it,' Lumia shouted at Abu Laith one afternoon, as he closed the gate to the courtyard behind him. 'I'm tired of waiting. I haven't got anything to do here. We're going out this evening.' She advanced on Abu Laith across the courtyard. 'And you,' she said. 'You'll be taking care of us. And if there's any trouble at all, we're going straight home, and there won't be any fighting.'

Abu Laith was slightly downcast. As the local *shaqawa*, or local hard man, he had always been the one called on to back his neighbours up in a fight, or enforce justice meted out by the local authorities. In this corner of Mosul, if someone bothered your wife, or stole from your shop, you knew who to call. Great fists clenched, belly protruding under his long *thobe* – a floor-length shirt – Abu Laith would storm up to your enemies, red-faced, and if necessary rough-house the perpetrator.

Lumia knew that this wouldn't work in the new Mosul. Some of her neighbours had been happy that the jihadis had come, and she didn't trust them. 'I mean it,' she said. 'We're going out, and we're all going to stay together.'

Abu Laith agreed, on one condition. They would go to the picnic area where families ate, and afterwards, if it was safe, they would drive up the river to go to see Zombie at the zoo. At that time of night during Ramadan, he thought, there wouldn't be many checkpoints to stop them. He was very excited to see the lion.

It took until late in the evening for everyone to be dressed and ready for the outing. By 8.30 p.m. Lubna and Oula, the oldest two daughters in the house, were still running around trying to get the smaller children dressed.

Abu Laith was, as usual, sitting on his sofa, watching the National Geographic channel, shouting out the best bits as they happened. 'Never split from the pack when a lion is chasing you,' he often cried, to no one in particular.

Lumia was never particularly impressed with his knowledge. 'Can you go and get the car started?' She barked.

Abu Laith was very proud of his Chevrolet truck, whose engine he had lovingly cared for since he bought it years ago, and which still ran perfectly. He bounced up and went into the courtyard.

Not long after, all the family were crammed into the spacious Chevy. The children fought to get their own seats in the back so that they wouldn't be thrown into the air when they bumped over one of Mosul's numerous potholes.

In the front sat the adults, Lumia's voice racketing away over the engine as she explained to the two oldest girls how she wanted them to behave around the jihadis. 'Don't look at them,' she said. 'These people are repressed. They've never seen a girl dressed normally before. If you look at them they'll try to marry you, so watch out.'

Oula, who was eleven and wore her hair flowing over her shoulders, a scarf vaguely slung over the back of her head, laughed at her stepmother, and left the scarf hanging. Her sister Lubna, whose three children were playing in the back, laughed too.

Lubna had been twenty-one when she had her first child with a hastily procured husband, a geography teacher the entire family had unanimously decried as a wimp. She had fought it for a few years before conceding that they were right, leaving him to move back to her dad's house, and sharing custody of the children. Now she spent much of her spare time giving herself elaborate makeovers and doing solo photoshoots. Maybe, she thought, she might find a new, less pathetic husband.

So far, it hadn't worked. But tonight her dyed-blonde hair was escaping from behind a loose scarf, and her eyes were ringed with kohl, leaving her lower eyelid standing out white. She had on a shirt and a pair of skinny jeans.

The sun was starting to set, and as they drove through the city the children bounced high in the back – Abu Laith veering through the traffic, stray dogs and piles of rubbish that had accumulated on the sides of the road. Every so often, as he always did, he shouted out a greeting to another driver or passer-by, slowing right down to shake their hand and gossip, oblivious to the honking and curses that swelled around them.

They drove past the rows of shopping malls that stood on the main street perpendicular to the river, past the cafes where families would sit, smoking narghile and drinking sweet, stewed tea. Children tumbling, they rounded the final corner and the Tigris spread out on their right – black water glimmering under the glow from the cafes on the banks, ropes of fairy lights strung across the willow trees like a cat's cradle. On the other side they could see the Old City, a mass of domes and arches that crowded on to the river in a wash of sand and pink. At this distance the houses, each window glowing with light, seemed stacked on top of each other like a Jenga set.

For hundreds of years, since before the Seljuks had ruled the city, travellers to Mosul had written of its beauty, and the fertility of its lands. Sacked by the Mongols and conquered by the Ottomans, the city survived and grew to the chaotic metropolis that it had become that summer in 2014.

As the family drove on, the lights of the riverside district called the Forest blared ahead of them in a swirl of reds and purples and greens. On their left was the thicket of trees that gave the area its name. They banked a mile-long stretch of road along the Tigris that exploded each evening into something of an amusement park: a rush of lights and tinny, wild jingles that blared again and again from the speakers on the fairground rides, or from boomboxes lashed to tea carts with pieces of wire.

The chaos would only die down during very late nights in the summer, when the choking heat stayed long after the sun had set, the battered roads sweating into the morning. Only the fishermen and the late drinkers came then – barbecuing their catch on small fires under the bridges, or sipping arak until the pink crest of sunrise rose over the citadel.

Abu Laith parked the car by stopping it in the middle of the road, and jumped out of the driver's seat, picking up armfuls of children from the back. 'Come on,' he shouted, as the women climbed out with a little more dignity. They unloaded the picnic they had packed in the back. Geggo, who was too young to remember having gone there before, gaped open-mouthed at everything.

They were going to a cafe by the Tigris that they sometimes went to with their cousins. While Abu Laith and Lumia liked it for the views over the river to the Old City, the children were intrigued by rumours that there would be a loud cannon which fired glitter into the river to signal the end of the fasting day. The zoo lay about fifteen minutes drive up the river, and Abu Laith thrilled at the idea that he might see Zombie later that evening.

Abdulrahman and Mohammed, two of the younger children, had raced ahead and were already in the cafe. Lumia was a little more wary. Once she had got out the car, she took her time looking around. She saw the carousels that had opened in the last couple of years. There was a man selling toasted sunflower seeds from a metal barrel with a fire under it.

All around them, people were milling about – women, like Lumia, in hijabs, shirts and skinny jeans. Most of the men, like Abu Laith, wore the thick moustache favoured under Saddam Hussein's regime or were clean shaven and in smart blazers and shiny shoes. It all looked, Lumia later recalled, quite normal. Heartened, she started passing the bags of food to the others. The air smelled of orange juice and roasting seeds.

'Let's go and sit over there,' she said, steering all the children she could reach towards a table in the gardens. The area was already filled with groups of families, all sat at tables or sitting cross-legged around their own thin tarps on the ground, spread with food on paper plates: a communal platter in the middle, rice and chicken made at home – bowls of hummus and moutabal on the side – or store-bought lamb kebabs dripping fat into their paper wrappings.

Abu Laith, beaming, rushed off to shake hands with vague acquaintances. As the kids played, Lubna and Lumia laid the food out on the table. 'Where are all the jihadis?' Lubna asked. 'Everything looks so normal.'

The iftar cannon went off, and the children cheered as the glitter fell into the slow-running water.

Soon, the entire brood was squeezed around the dish of dolma that Lumia had made at home, shoving each other out of the way for a piece of bread, upsetting the cold cans of Pepsi that Abu Laith had bought from the waiter. The barbecue they had built smoked on the ground next to them. When it was hot enough, they would cook kebabs on the coals.

'See,' Abu Laith said to Lumia, amid mouthfuls of rice. 'I told you it would be fine. The jihadis will be too fixated with themselves to bother us.'

Lumia wasn't sure. A quiet was descending over the riverside, and an uneasy feeling was rising. People close to a pavilion about a hundred feet away were standing up, looking over towards the structure, and sitting down quickly again.

'What's going on over there?' Lubna said, and Lumia shushed her. She had seen the armed men in their Kandahari shirts walking out of the pavilion. She looked inside the cafe they had come from, and saw more of them sitting at a table, having dinner. There were women with them – dressed as she was in a hijab and jeans, carrying handbags. They looked foreign. Some of them were blue-eyed and white skinned. The others looked Central Asian.

'Don't look at them,' she hissed, scrambling to grab the children. Abdulrahman and Mohammed were still running around the picnickers nearby, oblivious to the change in atmosphere. She pulled Geggo over and sat him in her lap, gathering the rest around her.

'What are we going to do?' she whispered at Abu Laith, who was stuffing rice and peppers into his mouth.

He looked up, supremely unconcerned, then started when he spotted the jihadis. 'Be quiet,' he said, sharply. 'Abdulrahman, Mohammed.'

The boys, who had been laughing, ran back to the table. Slowly, the chatter faded and the Forest road grew quiet but for the circular, clashing jingles of the fairground rides.

Lumia saw, to her mounting concern, that Abu Laith's face was set with anger. She knew, she recalled later, that her husband would not sit back and listen if the fighters insulted her and the children. Lumia tied the loose scarf tight around her head, hiding the hair that flowed from the front – gesturing for Lubna and Oula to do the same.

'Cover up now,' she said, in a flat, furious tone that children and adults alike had learned to fear.

Abu Laith was staring at the fighters. Three had broken off from the main group, and were walking directly towards them. They were all young men with dark hair and long, wispy beards. They were dressed in long black or brown shirts and ankle-length trousers, and carried guns. As they walked, they smiled and greeted everyone they passed. Occasionally, they would stop to talk to a family group. Everyone always smiled back. Some of their women were walking with them, and they were smiling too.

Lumia didn't need to look to know they were coming. Wordlessly, she mouthed discouragement at Abu Laith, who was stewing with indignation.

The air crackled, and a voice came from the speakers that usually played music into the garden. 'Attention,' a male voice announced. 'Attention. The use of mobile phones is banned here.'

Across the cafe, families scrambled to put their phones back in their bags. Fury growing, Abu Laith watched the fighters as they drew closer. It made him angry to see everyone crawling for them – how they all accepted their rules so quickly. It was low, he thought, and weak too.

'Assalamu aleikum,' one of the men cried out, when they were just a few feet away. He was wearing a brown shirt and had black hair.

'Wa aleikum assalam,' said Abu Laith, as loudly and confidently as he could. He was not, he thought, going to let them think he was scared. Lumia and Lubna muttered greetings in return. The young man was, Lumia thought in a strange moment of realization, extremely handsome.

'How are you? What's your news? How are things? Are you good?' the fighter asked politely, using the barrage of greetings that preceded every interaction in Mosul.

'Alhamdulillah,' pronounced Abu Laith, praise be to God.

'Alhamdulillah,' the fighter said. There was something fixed about his smile. His eyes flickered to Lubna, and snapped to the ground straight away. He did not acknowledge the presence of the women.

'Brother,' he said, still smiling at Abu Laith. 'Praise be to God, we want everyone to live a peaceful and a holy life. We hope that you are not frightened by our presence, brother, and that you understand we want only to let you live a simple life.'

Abu Laith spluttered. He glanced at Lumia, who later remembered feeling half-dead with fear. She shook her head.

Abu Laith was barely keeping himself together. If only, he thought, he could punch the jihadi in the face. He would never patronize anyone again. But there was no chance he and his family would make it out alive. The other jihadis were walking towards them now, where they sat at their table. He settled his face into what he thought was a pious demeanour. 'Of course,' he said. 'May God protect you.'

The fighter smiled benevolently at him. The little shit, Abu Laith thought, furious at everyone and himself.

'Thank you, brother,' said the fighter, walking away with his friends.

As soon as they were out of earshot, Lumia scrambled to pick up the children, throwing the food into a pile in the middle of the tablecloth and grabbing it by the corners like a sack. Around her, people were doing the same. 'Hurry,' she whispered. Abu Laith looked more furious than before, and got up quickly, as if making to follow the fighters. Lumia, Lubna, Mohammed and Oula threw themselves on him simultaneously. 'No, Baba,' said Mohammed, sternly. 'We're going home.' His pride injured, Abu Laith picked up the quiescent children and allowed himself to be dragged towards the car. Lumia babbled the whole way. Everyone else was quiet.

On the way out, they walked past the cafe owner, who was standing outside the main door, beaming at everyone who passed him. 'Why are you leaving?' he asked them. 'Nothing will happen to you. You can stay here, it's safe.'

It was too much for Abu Laith, who had been tested beyond endurance, and finally snapped. 'Why are you letting these people in here?' he shouted, lunging at the owner, who jumped out of the way. 'This is a place for families.' Crisis was only averted by Lumia, who shoved herself between the men and led her husband off. The Daeshis were crowding into the road now, and the way towards the zoo was closed ahead. They would not, Lumia impressed on a struggling Abu Laith, be able to go and see Zombie. Downcast, he agreed.

'That's it,' Abu Laith said, when they had all piled into the car and pushed it down a hill, so that it would start. 'That's the last time we're all going out together until these people have gone.' And for the first time in the history of the family, all of the women nodded in agreement.

8

Imad

1971

There was, Imad thought, very little point in going to school. He didn't want any new friends, and the lessons were boring. His friends were outside, waiting for him in the street. Bobby and Jonny might have grown at an alarming rate under Imad's largely meat-based diet, but they were devoted to him.

Together, they were the kings of Mosul's streets. When Imad bunked off school, as he did almost every day, they would roam the city from Zuhour in the east to the alleys of the Old City, where dusty stone passageways were filled with men selling cardamom coffee and salty cheese. They raced through the green lands that banked the eastern side of the river, with Imad rarely winning. Sometimes – and these were almost the best times – people would try to fight them. The children at school thought Imad was strange. He wasn't interested in working or studying. All he wanted to do was to spend time with animals.

But if they wanted to fight, he was ready for them. He was

big for an eleven-year-old: his arms were already muscle-clad and his hard stomach covered in a solid layer of kebab-fed fat. A few times, groups of boys thought they'd take him on. It was a bad idea. Once Imad, Bobby and Jonny had ground their opponents into the streets, they would run around the back lanes of Mosul, hooting and barking with victory.

At school, Imad's classmates had grown used to his incessant bragging about Bobby and Jonny. They had seen the dogs wait for him, tails wagging, at the school gates – Bobby a yellow white, like a smoker's finger, and Jonny, much whiter with one brick-red foreleg – and had seen how they burrowed into his face when he ran out to meet them. As they talked in the playground, Imad's classmates agreed that the dogs seemed quite friendly.

Imad had laughed at the suggestion. 'They are strong,' he had told his friends. 'They are fierce. You just try to hit me when they're there, and you'll see what happens.'

Jaffer and Auni didn't believe him. They were cousins – new to the neighbourhood and determined to make their names. Both came from the Surchi tribe, which had an unfair reputation for being slightly dim and beholden to their sheikh. They wanted to carve a reputation for themselves, and the only way to do it was to break Imad's eyes, as the saying went, by beating him.

One afternoon, they waited on the street outside a restaurant in Zuhour for Imad to walk past with his dogs. They came, as usual, looking inordinately pleased with themselves. Jaffer, who was taller than his cousin, with blue eyes and a mean stare, took the initiative and lobbed a stone at Imad's head.

Imad's world went black, and before he could react, Jaffer bounded forward and pushed him over, almost pitching him into the stream of dirty water that ran from the restaurants down the side of the road.

The ground was spinning around him, and Imad could barely tell which way was up. But he could hear the barking and the shouts as Bobby and Jonny tore into Jaffer's clothes, pinning him onto the floor and savaging him thoroughly. His friends were saving him. Head lightening, Imad sprang up and went to work on Auni, who was standing in shock by the side of the road. By the end of the day the story of Imad's murderous dogs, who mauled anyone who got near them, had spread through the neighbourhood, much to Imad's satisfaction.

That evening, as they sat eating dinner on the front step of the house, Imad's mother came over, looking very angry. 'You've done it now,' she said, and pulled him up the stairs to his grandmother's room.

Inside were two other women, who looked just as angry as his mother. His grandmother, however, looked close to laughing. 'I hear you've been fighting,' she said. 'Auni and Jaffar's mothers are here.'

'My son has a cut on his hand from your dog,' spat one of the women.

Imad glared at her. 'He shouldn't have tried to beat me then,' he said. 'They were only protecting me.'

The sons were sent for. In the bedroom, under the stern eye of the women, they were forced to shake hands with Imad, as Bobby and Jonny waited outside in disgrace. The boys glared at each other, thinking of revenge.

By the next day, all was forgotten. Together Auni, Jaffar, Imad and the dogs were running wild through the streets of Zuhour, stoking chaos wherever they went.

9

Hakam

Two weeks after the jihadis took Mosul, Hakam Zarari left the house to go and have a look around the city. There was no point, he reasoned, in waiting at home for things to change. Outside the window on the landing, the city seemed quiet. Since the militants had arrived there had been no car bombs, and he was bored. Hasna, his sister, who was on her summer holidays, needed to stay at home anyway to revise for her exams that autumn – by which time, they assumed, the jihadis would have gone.

In the thin light of the morning sun, Hakam got into his car and drove towards the city centre. The streets were emptier than usual – both of cars and of people. From habit, he drove on the only road that the government had left open in his neighbourhood. When he came to the junction where the checkpoint usually stood, he slowed down.

There was no one there. The queue that always stretched for half a mile or so into the neighbourhood was gone, and the checkpoint destroyed. Someone had pulled aside the blast barriers that usually blocked off all but a car's width of the road, bringing traffic to a standstill. A pile of burst sandbags marked where the machine-gun emplacement usually stood, guarded by at least four soldiers.

Hakam turned the corner on to the main road. A few of the shops were open, spilling their goods on to the street. Vegetable sellers had brought out their carts, and were standing in the ditch as usual, surrounded by rotten cast-offs and peelings.

As he drove, Hakam felt a tiny thrill. He could drive wherever he wanted. At this rate, it was going to take him ten minutes to get to work, rather than the usual hour and a half. He had nothing in common with the jihadis, but the government hadn't treated him well. These guys were good administrators, he thought. They knew the way to the people's hearts. Stop the traffic jams, and the rest will follow.

Up ahead, he could see the remnants of another checkpoint that had been cleared from the road. A group of men were standing next to it dressed much like he was, clean shaven in jeans and t-shirts, but with Kalashnikovs hanging by their sides. Hakam slowed down as he approached, ready to stop the car. He was afraid.

But as he drew closer, he saw that they were waving at him to pass, grinning with an air of maniacal cheerfulness. 'How are you? What's your news?' they shouted as he passed.

Hakam, who didn't quite know what else to do, smiled back and drove on, relief flooding his veins. They must be trying to get us on their side, he thought, as he drove on towards a friend's house. It was odd, but it wasn't bad compared to the army. For most Moslawis, these men were just the latest in a bewildering array of occupying forces. And unlike the last lot, they didn't block off the roads.

While Hakam was driving around Mosul, the militants – many locals or fighters from the other Sunni provinces to the south – were busy working. They had fought the Americans, then the Iraqi army. Now they controlled Mosul, Iraq's northern Sunni powerhouse, after barely even a fight. They were not going to squander the opportunity, and they knew they had to get the local population on side.

Across the city, they destroyed the barricades that had inconvenienced Moslawis. They pulled down the checkpoints where young and old alike had been humiliated. In a city that had been strangled with military rule and the constant threat of arrest or bombings, calm was descending for the first time in years.

People started to visit long-lost friends who lived on the other side of town. They flocked to markets without fearing the regular Friday attacks. With the suicide bombers in charge, and the army gone, the streets felt decidedly calmer and safer.

The fighters didn't seem to mind that men like Hakam wore skinny jeans, smoked heavily and believed, above all, in science. Nor that Hasna walked around in trousers, a long shirt and a hijab, her face uncovered. In those first few weeks of their rule, she wore the same clothes she always had, and the jihadis said nothing about it. They affected not to care that Hakam's father Said had insisted on following the letter of the secular Iraqi constitution, as long as he stopped doing it now.

On other things, they were less indifferent. A few weeks after the militants had arrived, Said had tried to get in touch with the Mosul court to see if it was open. It wasn't: the jihadis had burned it down because they thought that any law other than God's was forbidden. Said had been so angry when he heard this that he hadn't quite known what to do with himself.

Whenever he left the house, Said seethed at the ignorance and blind belief that dominated in the new Mosul. One afternoon not long after the militants arrived, he went to the bank – straight-backed and imposing, as always, in his astrakhan hat. It was best, he reasoned, to take the family's money out before the jihadis realized that they could fund their own operations by stealing it.

The jihadis had already taken over the bank, replacing the tellers with their own men. Said looked down his nose at them as he strode into the dusty building. They were ignorant, he thought, and purposefully so. Said was deeply religious, and very proud of his faith. He wanted nothing to do with these people who claimed to speak for his God.

The teller looked over Said with interest as he made his withdrawal, taking in his bearing and his well-cut clothes. 'What's your job?' he asked.

'I really don't think you want to know,' said Said, his tone icy. He didn't want to start a discussion with these people if he didn't have to.

'No, go on,' said the teller. 'It's fine, whatever it is.'

'Fine,' said Said. 'I'm a lawyer.'

The bank teller sneered. 'That's *haram*,' he said, confidently. 'The Iraqi law isn't God's law. It's the law of humans, and you shouldn't follow it.'

Said's anger had risen before he could even think. 'Oh really?' he said. 'And where do you think Iraqi laws come from?'

The teller looked confused.

'I'll tell you,' Said said. 'Most of Iraqi laws, and the Iraqi constitution, come from sharia law. If you look at the laws of commerce that govern what this bank can do, it is almost exactly the same as the one that is written in sharia. That's because it's fair.'

A crowd of tellers had been listening to the conversation, drawing a little closer as Said lectured them in the decisive, clipped tones that came from years of speaking in court.

'I didn't know that,' one of the tellers said.

'Do you want to come and work with us?' the teller who had served him asked. 'We could use someone like you.'

Said had to try very hard not to shout at them. He was disgusted that they might think, even for a moment, that he was one of them.

'No,' he said. 'Absolutely not.' Taking his money, he strode out of the bank.

One Friday soon after, Hakam was sitting upstairs in his bedroom on Facebook, scrolling half-distracted through posts about protein supplements and cat videos. On and off, he kept up the chat windows he had open with friends from across the world – gym bros in the States, girls in Germany. He'd met them through Instagram, forums or – back when the internet was a simpler place – Yahoo chat. The internet hadn't been affected by the takeover, but he hadn't spoken much to his friends about what had happened. The ones that he had told him to leave, but he told them he couldn't because of his family. Most didn't know where he lived. They talked about other things.

Downstairs, his parents and Hasna were watching TV in the living room. The Middle East had been slipping down the news cycle over the last week – the focus was now on student protests in Hong Kong.

Then his mother shouted. 'That's in the al-Nuri mosque,' she said. 'He's in Mosul.'

The al-Nuri mosque in the Old City famous for its leaning minaret, known to locals as al-Hadbaa, the hunchback. It was commissioned in the twelfth century by Nur al-Din, a Seljuk sultan known for his ultimately unsuccessful attempts to unite a Sunni army against the Crusaders. But if there was anything that brought together the people of Mosul, it was their love for al-Hadbaa, the jaunty-angled cornerstone of their city.

Hakam ran down the steps of the swerving staircase and into the living room. A man in a black turban and long robes was on the TV, leading Friday prayers in al-Nuri. The family sat on the sofa, listening to every word.

'I am the leader who presides over you,' said the man, as a black flag logo waved in the corner of the screen. 'I was put to the test by this heavy responsibility with which I was entrusted.'

In front of him were a crowd of men, seemingly ordinary Moslawis, praying. They looked scruffy and uncomfortable, glancing occasionally at the camera.

'Who is this guy?' Hakam asked Hasna, who was gaping at the TV.

'No idea,' she said. 'But he says he's the caliph.'

There hadn't been a caliph, or an Islamic ruler, since the Ottoman empire collapsed in the 1920s. For centuries, the sultans had ruled large swathes of the Muslim world from Constantinople – an empire that at one point stretched from the gates of Vienna to the warm blue waters of the Persian Gulf. These sultans were caliphs, the representatives of God on earth.

Still earlier, from the eighth century, they lived in Baghdad, where they were the rulers of the Abbasid caliphate. When Europe was scrabbling through the Dark Ages, the Abbasid rulers were overseeing highly advanced studies in geometry and astronomy. They kept some of the finest libraries on earth, full of Greek texts saved from the destruction that befell European culture after the collapse of the Roman empire.

This legacy, over a millennium of Islamic learning and empire, was contained and implied in the title of caliph. As Hakam watched the man on the TV speak in his elaborate formal Arabic, face pale under the harsh fluorescent lights of the prayer hall, he felt like laughing. The whole thing seemed utterly ridiculous.

'Who chose him?' snapped Said. 'A leader should be honest and good. Who chose this man?'

Hakam stared at the worshippers. They looked local. He doubted they'd known about this new caliph either. They'd probably just gone to Friday prayers and been faced with this bizarre distortion of history. It was odd, but it didn't seem particularly threatening.

A month or so later, as the summer of 2014 dragged on, a message went out that government employees should go back to work. They would be paid their salaries by the government in Baghdad. Hakam was pleased: he had been bored out of his mind and broke. With his colleagues around him, he settled back into life at the lab: driving to work in the morning through the open roads, going to the gym, hanging out with his friends.

In the barber shops across the city, men had their beards shaved. In the cafes, they smoked cigarettes and drank tea. Women went shopping wearing trousers and shirts.

Life was very much as it had been before.

10

Abu Laith

WHENEVER LUMIA LEFT THE HOUSE ON VISITS TO HER relatives or to the shops she followed the same routine. First came the bone-white face powder a good three shades lighter than her skin. On went the kohl, which crinkled pleasantly into the corner of her eyes. Eyeshadow, often in an arresting pastel, came out for special occasions. She would put on a long dress, often a blue velvet number sprinkled in sequins, and march around the streets, shouting her hellos, with her scarf wound loosely around her thick black hair.

Lumia was proud that she was beautiful, and very happy to show it. She was not going to listen to her sister-in-law's stupid ideas.

'You're going to have to start wearing it,' she told Lumia one afternoon, holding out a bolt of thick black chiffon. 'I told you.'

Lumia was feeling extremely truculent. 'I'm never wearing it,' she said. 'It's horrible.'

A few days before, the militants had declared that women must cover their faces, and that their husbands and brothers would be punished if they didn't. Few had followed the rules.

Most had just ignored them and gone about their business. But now signs had gone up around the city, pasted on walls and signposts, warning that women should dress in long black robes, with a cloth covering their face.

'I just got stopped at a checkpoint,' her sister-in-law had said, pale and worried, when Lumia had dropped by her house that afternoon. 'They told me I had to cover my face and my eyes.'

'Are you kidding me?' asked Lumia. 'How are they going to make you do that?'

To many Moslawi women, the idea of wearing the face-covering *khemmar* was bizarre, almost funny. While most Muslim women in Mosul wore the hijab in public, covering the face was something done only by Wahhabis in the Gulf, by villagers in the desert or the ultra-religious. In Mosul, two types of women wore the *khemmar*: the unusually devout, or the unusually promiscuous who wanted to conduct their love affairs in secret.

Lumia was not, she said, going to wear something that made her look like a mullah's wife or a prostitute.

'It doesn't matter what you think,' said her sister-in-law. 'They won't punish you, they'll punish Abu Laith.'

That made Lumia pause. She had spent her whole life battling attempts by men to control how she spoke or what she did. As a teenager, she had turned down a marriage proposal because her intended wanted her to wear the *khemmar*. But all of that paled into nothing compared to the stark horror of what would happen if the jihadis took her husband.

'Fine,' she said. 'I'll wear it.'

The two women spent the afternoon at the sewing machine, running up a three-piece, one-size-fits-all ensemble. First came the abaya, a long-sleeved, floor-length, neck-high robe that did up at the front like a coat. Then Lumia tied her hair into a bun and covered it in a hijab, holding one end of the scarf tight under her chin and wrapping the rest around her head so that it sat snug over her neck and hair. Then, with two squares of material sewn together, came the *khemmar*, which covered all of her face apart from her eyes, and tied at the back of her head with tight elastic that hurt her head. It fell down over her shoulders and chest to her knees, like a second abaya that also covered her head.

At this point, only her hands, eyes and feet were visible. Her sister-in-law took the final part of the ensemble – a large square of black cloth – and draped it over her head, tucking it into the back of the *khemmar*. Lumia's breath felt hot under the layers of cloth. Her sister-in-law was looking at her, and she wasn't laughing, and neither was Lumia, who, she later recalled, felt like she might faint.

On the way back home, she stopped at the market and bought a pair of black gloves and some thick black socks. When she wore the gloves, she couldn't count money, and kept dropping it on the floor. Worse, she could barely see, and when she got in and out of the taxi, she tripped on the hem of her abaya. Everywhere there were women just like her, swathed in black and walking unusually slowly.

By the time she got home she was very angry, in particular at Abu Laith, who had been moaning about his lion again. He hadn't been able to go and see Zombie since the jihadis came. The road that led to the zoo wasn't safe. The militants had – he heard – opened a training camp right by the zoo, and civilians were avoiding the area. He couldn't see a way to go there without risking being arrested. No one, he later said, could go there without running into the jihadis. He hoped that one of the workers at the zoo would keep feeding Zombie.

'If the jihadis touch Zombie,' Abu Laith had kept fuming, to no one in particular. 'I'm going to find out who did it. And they're going to regret it.'

Even if Lumia ignored him, he would continue.

'If I could just hear a roar, I would know if it was Zombie or not,' he lamented. 'I would know him from a thousand lions.'

Lumia hadn't been home long when there was a knock on the door. Immediately, as ever, a wave of skinny-limbed children and assorted animals came rolling up to the gate, fighting to answer it.

'Who is it?' they all shouted. 'Who is it?'

'Abu Rafal,' said the voice outside the gate. 'I need to talk to your father.'

Abu Laith, who had been following the commotion at the gate with interest, walked over and opened it. He was not particularly looking forward to seeing Abu Rafal, who he thought was a bit of a wimp, but he answered anyway.

'Assalamu aleikum,' said Abu Rafal, when the gate swung open.

'Wa aleikum assalam,' said Abu Laith.

The two men exchanged polite greetings before Abu Rafal spoke up. 'The *dawla* have come to the mosque,' he said, using the word – state – that the jihadis used to describe themselves. Those who didn't support them said Daesh, an acronym of the group's full Arabic name. 'They're asking about you. They've been wondering why you haven't been praying.'

Abu Laith could see the turquoise mosque behind Abu Rafal, a one-storey building with its spindly minaret. Abu Laith, in a fit of piety, had paid for the mosque to be built with the money he saved from fixing expensive cars over a decade ago, but had stopped going after he realized that most of the worshippers just came for the free breakfast he gave out at morning prayers. Though he remained close to a few of his old pious friends, he had grown furious at the hypocrisy he saw in the mullahs who grew rich and supported al-Qaeda while pretending to be God-fearing scholars.

Now, after a few drinks, two-faced mullahs were the favoured subject of his rants, along with interfering relatives and people who beat dogs. His religious friends, who were used to it, just grumbled and admonished him.

Abu Rafal, however, was not a friend, and certainly not someone who could order him about. 'I'm not coming,' said Abu Laith. 'They're all hypocrites. They pray during the day, and in the night –'

Abu Rafal was now very agitated. 'You have to come,' he said. 'We were in there just now and they were talking about how they have to move some things from the mosque. I told them that they had to ask you first.'

'Exactly,' said Abu Laith. 'Because I built this mosque, and with my own money too.'

'I know,' Abu Laith later remembered Abu Rafal saying. 'But there's a Daeshi called Abu Hareth. He's moved into the house those Kurdish people down the road left behind. He's a big guy in Daesh, and he was asking for you. He said he'd heard your name before and wondered why you weren't going to the mosque. He wants you to come right now.'

Abu Laith was indignant. 'A Daeshi?' he shouted. 'Order me to come to the mosque that I built? Who does he think he is? I'll go there when I want, and you can tell him that.'

Uncharitably, Abu Laith slammed the door in Abu Rafal's

face and marched back into the house. 'I told that mullah-worshipper where to go,' he crowed to Lumia. 'I said I built that mosque with my own money, and that they couldn't boss me around.'

Lumia, who had heard a version of the mosque-building story at least once a week for the entirety of her married life, stopped listening. She had long ago resigned herself to the fact that her husband's big mouth would get them all into serious trouble.

One evening not long afterwards, Abu Laith was sitting on his bed on the first floor, looking out of the window at the street below. The lights were on in the mosque, and he could see a steady stream of people going inside for evening prayers. Then it went dark.

'They've burned the bloody generator out,' shouted Abu Laith, consumed with rage. 'I paid for that generator.'

Abu Laith, though he sometimes struggled to remember exactly when his children were born, knew to the dinar how much he had paid for everything he owned, and when. The generator, Abu Laith reflected, as he stormed down the stairs, through the courtyard and into the street, had been extremely expensive, and as such could have been reasonably expected to last at least another five years, if not more.

He raced up the stairs to the mosque. He had grudgingly tolerated these people coming to his city and making his wife wear a blanket. But he was not going to have them destroying his property on the pretext of some hypocritical holier-than-thou ideology. They were going to have a talk, man to man, and they would see sense.

'May peace and mercy and the blessings of God be upon you,' said Abu Laith as he strode into the mosque. It was the most pious Islamic greeting he could think of, and he knew it would make the Daeshis take him seriously.

'Wa aleikum assalam,' one of them said, turning to face him. He was large, not fat exactly, but a solid mass of belly with a stoutness that made him look immovable. His accent sounded Moslawi, but Abu Laith didn't recognize him. He had pale skin and a long beard, and he looked overwhelmingly confident.

'You've burned out the bloody generator,' Abu Laith shouted, advancing on the man. 'I paid for that generator, and I built this mosque with my own money, so I'll thank you not to ruin it.'

The man looked at him. Next to him stood a large blond man, who didn't look as if he understood what was going on, and a few locals.

'If you built this mosque,' said the pale-skinned Moslawi, 'why do you never come here to pray? I know who you are. You're Abu Laith. We've heard about you. We know you drink alcohol.'

Abu Laith was not listening. He had noticed that the Islamic blessing inscribed in swooping calligraphy on the wall had been ruined – painted over very recently, by the look of it. 'You've destroyed my mosque,' he said. 'What have you done? I paid two thousand dinars for that in 2012.'

The Daeshi still looked calm. 'It's *haram*,' he said. 'It is like a picture. God shall punish any who gazes upon his image.'

Abu Laith swung round. He was angry about the generator, and angry about the calligraphy, but the sheer gall of the Daeshi to come to his mosque and lecture him about the Quran was so biting that he wanted to scream. 'If this is your Islam, screw your Islam,' Abu Laith shouted. 'I don't want anything to do with any of it.'

He turned around and marched out of the mosque, cursing as he went.

The next twenty-four hours in the house were unbearably tense. Lumia was angrier at Abu Laith than she ever remembered. She could not understand – she later said – why he insisted on provoking everyone and putting the family in danger. The second reason was that Abu Laith – who was not listening to Lumia – had taken up a new post by the front door, where he was keeping guard over the mosque.

Since he had left in a huff, he had suspected that the new mullahs were planning on stealing the things he had bought – at great personal expense – to furnish the mosque. It didn't matter that Abu Laith never went to the mosque itself, and disliked most of the worshippers. This was a matter of principle. 'Make sure you check the air conditioner is still there,' he had shouted at the worshippers streaming in for afternoon prayers. 'Otherwise they'll probably steal that too.'

They didn't. But they did come knocking on the door the next day around the same time, looking for Abu Laith. Lumia ushered the children into the back room, where she threatened them into silence. She locked the door that went through to the living room and sat there, hoping that this time Abu Laith wouldn't do anything too stupid.

'Welcome,' said Abu Laith when he opened the door to the visitors. The large blond man from the mosque was there, as was the imam and a few of his acolytes. The large, pale-skinned Moslawi was nowhere to be seen. They filed in, muttering religious greetings and looking around suspiciously at the courtyard, which was as always scattered with toys and fruit peelings. All but the imam had guns hanging from straps on their shoulders.

The blond Russian – Abu Laith thought he was Russian – spoke strange, broken, formal Arabic, and looked decidedly out of place. 'Thou art needed to complete the prayers at mosque five times a day,' he said, in a scholarly tone, as they stepped into the living room. 'Thou art commanded to not consume the alcohol drinks.'

'You haven't even sat down yet,' said Abu Laith, 'and you're already telling me what's *haram* and what isn't.'

The imam and his friends looked uncomfortable, but the Russian ignored him. He was looking across Abu Laith's living room, with its beige clinker-tile floors and brown sofas, to the walls on the opposite side of the room. 'Thou art depicting the Prophet, peace be upon him, in his own image,' said the Russian. 'Thou knowest these images are forbidden.'

He was pointing at the pictures that hung on the far wall: one of Abu Laith, young and wearing a red and check kaffiyeh scarf on his head, and one of his brother, who looked almost exactly like him, and had been martyred in the Iran–Iraq war.

'Are you kidding me?' Abu Laith later remembered saying. 'That's not a picture of the Prophet.'

But the Russian was striding over to the wall, and had jumped on to his sofa – bought at considerable expense ten years ago – to take the pictures down.

Abu Laith was momentarily dumbstruck. He had always said that Mosul was like a camel – it pissed backwards, not forwards. But this was something else. This was cultishness of a kind that he hadn't seen before.

But now the pictures were down, and the Russian was

staring at the Quranic citation written in swirling calligraphy and moulded on the wall in plaster. Without saying anything, the fighter started pounding the inscription with the butt of his rifle.

Both the imam and Abu Laith were gaping at the Russian now. Shards of plaster flew everywhere, and dust filled the air.

'What the hell are you doing?' shouted Abu Laith, but there was no point. The man seemed to inhabit a different world.

When he was done, he stood there, panting, and turned to face them. 'Thou will to go to the mosque five times a day,' he said, and marched out the door, his companions trailing after him.

Abu Laith followed them out into the garden and watched them leave. 'Not a chance,' he considered, closing the gate behind them. 'These people are mad.'

11

Hakam

THE MARKET BY HAKAM'S HOUSE WAS HEAVING, AS IT ALWAYS was in the afternoons, with a sweaty, cantankerous crowd of Moslawis going about their daily business. Perfume from the soap stall drifted in the air, and mixed with the reek of the rotting vegetables crushed into the lee of the kerb.

In the afternoons, before they went to the gym, Hakam and his friends would drift through the market, calling out greetings and occasionally buying something. With the new rules that Daesh had imposed there wasn't much else to do.

That day, Hakam had bought a pair of trousers at a market stall. They were black skinny jeans, and he carried them in a plastic bag.

Mohammed, a friend from school, had a shop on the corner. Hakam was about to walk inside when he saw Abu Jaffar and Abdulkareem across the road. He had to suppress a smirk. The two men were hisbah – religious police appointed by Daesh – who patrolled this market. They were also the butt of the entire neighbourhood's jokes: village boys from the outskirts of Mosul whose combined intellect rivalled that of a flea's, and whose crime-fighting powers were not much better.

Their brutality, sadly, was more highly developed. These days, people were going missing: picked up by the Daeshis for some minor infraction and never seen again. That kind of thing had of course been happening in Mosul for a long time. But before, kidnappings were a much bigger threat than being arrested. If you had money, you could get free. Now, just a few months after the militants had arrived, you could just vanish.

The friendly face that Isis had put on when they first arrived was slowly slipping, and now there were stories of killings and a constant unease. When Hakam saw the two hisbah members, he crossed the street and hoped for the best.

Abu Jaffar and Abdulkareem relished their power, as minor tyrants do. That afternoon, they strutted down the street, their long hair hanging down over the scarves they wore around their necks. Their uniforms were black vests with 'hisbah' printed on the back.

'Hey, you,' Abu Jaffar shouted, pointing at a teenage boy who had rounded a corner in front of him. 'What have you done to your hair?'

The young man stopped, his hand flying to his hair – a slick black creation combed with gel. He looked extremely guilty. Groomed hair that looked like it belonged on a Turkish soap star definitely fell under the rubric of things that Daesh disliked, though they had not got around to banning hair gel explicitly yet.

'Nothing,' he said.

Abu Jaffar and Abdulkareem walked towards him. 'You've been caring for your hair like you're a girl,' said Abu Jaffar. 'Why are you making yourself pretty?'

The young man was looking around for an escape route. Hakam's friends had gone into the shop, but he kept watching. They hated seeing good-looking men, he thought, taking in the hisbah members' black tresses curling greasily on to their shoulders.

Abdulkareem was still shouting at the boy. But Hakam noticed that Abu Jaffar was staring at him. He made to turn around and go into the shop.

'Hey you,' the fighter shouted, this time pointing at Hakam. 'You, with the glasses.'

Without really considering what he was doing, Hakam opened the door of the shop and walked inside. He could hear running steps following from behind. His friends were inside the shop, joking with Mohammed. They looked up at him for a second, before the door clanged open and Abu Jaffar came into the shop.

Hakam turned around. The hisbah member was looking deeply pleased with himself. 'Hey, you,' he said. 'Your trousers are too long.'

The militants' latest rule had only been announced a week ago, and Hakam had been hoping they wouldn't enforce it. An edict circulated in the markets had stipulated that trousers must be worn above the ankle bone, as the Prophet Muhammad had (allegedly) done.

At six feet tall, Hakam found most trousers too short. But wearing ankle-swingers made him look like an idiot. Was he meant to cut the bottom five inches off all his trousers because the hisbah thought it was a good idea?

Smiling at Abu Jaffar, he held out the shopping bag he'd been carrying with him. 'Oh, I know,' he said, pleasantly. 'That's why I went to buy these new trousers.'

Abu Jaffar grabbed the bag and looked inside, before pulling out the jeans. He looked confused for a second, then angry. 'Hey!' he shouted. 'But these are long trousers!'

Hakam was enjoying himself. 'What?' he said, bewildered. 'I don't understand. They looked shorter in the shop. Maybe they gave me the wrong size?'

Abu Jaffar paused for a second, looking at Hakam as if he suspected a trick. But Hakam assumed an expression of confused outrage, before taking the trousers from him. 'I'll have to take them back to the shop,' he said, trying to sound dejected. 'I came out to buy some halal trousers as soon as I heard the edict. Forgive me, but I didn't know beforehand. It's not my fault the people in the shop gave me the wrong pair.'

Abu Jaffar looked at him, not quite sure what to do. 'Show me your ID,' he said.

This was not good. If they saw Hakam's ID card, they would be able to find him again. They could go to his house, bother his family, go to his workplace. Under rules in place before Daesh, everyone had to carry their national identity card at all times.

'I don't have an ID,' Hakam, later remembered saying. 'The cards were given out by the infidel government. It's at home, and I was planning on burning it. I've been waiting for the Islamic State to make their own IDs, for the true believers.'

Abu Jaffar looked more perplexed than ever. 'Where's your home?' he asked. 'Let's go and get it.'

'It's really far,' said Hakam, vaguely, but his apprehension grew. 'Ages away.'

The militant lunged forward, pointing at one of Hakam's friends, who were all looking nervous, staring into the dark corners of the shop. 'How about I take him, then,' he said. 'I'll keep him here until you come back with your ID.'

Then the door crashed open and Abdulkareem swept in. Hakam hoped it meant that he had let the young guy with the fancy hair go. 'What do we have here?' he asked.

Hakam tried to explain yet again about his new trousers, his ID, the immense distance to his home. Both the Hisbah members looked as if they were about to hit him. But as Abdulkareem was about to speak once more, he caught sight of something out of the shop window. 'Wait,' he said. 'Is that woman not wearing gloves?'

Abu Jaffar turned to the window. 'She's not,' he said.

The two of them rushed towards the door. 'If I see you wearing infidel trousers one more time, we'll take you in,' Abdulkareem said, in a parting shot.

Hakam agreed. 'Don't you worry about that,' he said, as they rushed out of the door towards the unsuspecting woman. He waited for the door to slam shut before he cracked up laughing at Abu Jaffar's obsession with infidel trousers.

He turned to his friends. None of them seemed to think it was funny.

12

Abu Laith

'That is it,' said Abu Laith one afternoon to two of his daughters. He was sprawled, as usual, in the living room, where he had – due to space restrictions and a lack of mirrors in the house – been forced to watch them put on make-up and take selfies for an unconscionable amount of time. 'You're going to have to leave.'

Lubna and Oula had known it was coming. Since Lumia had first put on her *khemmar* at her brother's house, the blanket laws, as Abu Laith called them, were growing more pervasive. The religious police had multiplied, and they roamed the streets looking for a flash of stray ankle or uncovered eyes. When Lumia went shopping, she was swathed in black, from her toes to the crown of her head, which was always covered in at least three layers of black polyester.

Abu Laith was not, he thought, as he looked at his daughters – who usually dressed in shirts and skinny jeans with sparkles on the back pockets – going to have them wearing that get-up. He hadn't let them go outside since the night of the picnic in the Forest. Though Daesh might pretend to be friendly, Abu Laith knew they were extremely dangerous. On the rare occasions he ventured out into the city, he avoided the main Daesh checkpoint at the end of their road. He had tried everything, called everyone he could think of, to try to figure out a way he could get to the zoo by the river to see Zombie. But for now, it was too close to the Daesh camp for him to be able to go safely. As far as he knew, no customers went there any more. They were afraid of the gunfire from the training camp where they were teaching the new generation of 'cubs of the caliphate' – the children of Daeshis – to fight. Day and night, Abu Laith worried about Zombie. Since the Daeshis had arrived, he hadn't been able to reach Ahmed, the zoo manager, on the phone. Abu Laith could only hope that Zombie was being fed, and wait until things had quietened down enough for him to visit the lion.

For his daughters, going outside in that early autumn of 2014 was in some ways far more dangerous than for the rest of the family. Unlike Lumia, they were young and unmarried. If Isis fighters saw them, there was a chance they would try to abduct them for their own use. So when his oldest daughter Dalal called to tell him what to do, he actually listened for once.

'Baba,' she had said, in her husky voice. 'You have to send them to me.'

Dalal lived in Baghdad, where she worked in Iraqi military intelligence. She was twenty-nine years old, efficient and bottle-blonde, with a blade-sharp determination. She didn't listen to anyone who tried to boss her around, including her husband, who had been ineffectually whinging at her for years to give up her job, and her father.

'I can't send them alone,' Abu Laith had said.

'Baba, it's urgent,' Dalal had said. 'You don't understand these people. It's going to get much worse.'

About this, Lumia had agreed. 'The jihadis are going to try and marry them even if they just see their eyes,' she insisted. 'And you won't be able to stop them.'

'They've taken women as slaves,' Abu Laith later remembered Dalal saying. 'They'll take the girls if they know they're not married. They want to have as many babies as they can to fill out the caliphate.'

Abu Laith, stuck between two forces, struggled. It was a risk to send the girls away. They could be captured on the way out and imprisoned or married off. But if they stayed in Mosul, they wouldn't be able to leave the house, and would live in constant fear of being forcibly married to some fundamentalist madman.

'How are they going to get to Baghdad?' Abu Laith asked. 'I can't take them. The jihadis won't let me back into Mosul if I leave. We'll lose the house. And I can't leave Lumia here alone.'

Dalal's voice was grim on the other end of the phone. 'Leave it to me,' she'd said. 'I'll take them.'

The danger in which Dalal placed herself by coming to Mosul is impossible to overestimate. Baghdad was six hours away by road, but since Isis had arrived it might as well have been in Mongolia. The checkpoints on the way were manned by suspicious jihadis. Men who left the city had to leave behind the deeds to their house as collateral. If they did not return, the house was given to a devoted member of the *dawla*. Women could not travel without a male guardian.

More than anything, the Daeshis were looking for spies, and for government employees, who had to sign a *towba* – a declaration of repentance and fealty to the Islamic State way – or face being arrested. Dalal was a woman, travelling by herself, who worked not only for the government, but for military intelligence.

Still, she was coming. Strangely, Abu Laith wasn't that worried about her. She knew what she was doing.

'Lumia,' Abu Laith said, in a decisive fashion, to his wife. 'You have to go with them. I'm not letting you stay here.'

Lumia was unimpressed. 'I'm staying,' she said, firmly. 'If you die, I die, and that's just how it is.'

'But the other kids will have to stay too,' Abu Laith said, thinking of Luay, his second-oldest son, and the three smallest children.

'Yes,' said Lumia. 'But they'll stay inside, and it'll be over soon anyway.'

It didn't take long before Dalal had made the necessary travel arrangements, and was driving up to Mosul from Baghdad, a distance of 250 miles. She wouldn't come to the house, she had said, in case she was being followed, but would meet the girls at her sister Ridha's place closer to the city centre.

Lubna was extremely relieved to be going. She could bring her children with her, and live without her relatives breathing down her neck, nagging her about her divorce. She could go outside without being concerned that a teenager with a gun would decide to marry her after glimpsing her eyes.

There remained the problem of the male guardian. Women were not allowed to travel alone, and Abu Laith and Luay had to stay behind to protect the house. There was only one other option.

'Mohammed,' Abu Laith announced. 'You're going to Baghdad with your sisters.'

Mohammed, who was thirteen and bored, agreed, without great enthusiasm. He knew that his sisters would be in charge, and that the guardianship was only in name. Still, he thought, Baghdad might be fun.

On the day of departure, the girls and Mohammed gathered in the living room and hugged their father and Lumia goodbye. They only had small bags with them, so they wouldn't attract attention. Dalal had found a smuggler who would take them out for $2,000. But he just had one small car, and there were seven of them, and the smuggler himself. They would drive through the desert, on a secret route that passed through Syria and down along the border before going back into Iraq. It was good, Abu Laith reflected, that Oula and Mohammed didn't realize how dangerous it was.

'It won't be long,' Lumia said to Abu Laith, as the children filed out of the door and climbed into a taxi. 'They'll be back soon.'

Once his siblings had left, Luay, the would-be geography student, found himself at a loose end. While he was quite as good-natured and strong-boned as the rest of the family, he didn't have Dalal's drive, nor his father's confidence. He spent much of his time playing games on his phone, or doing the occasional spot of manual labour. Since Daesh had arrived, he had been subjected to lengthy rants from his father about Zombie, and his fears for the lion's health. Together they had worked to try to find a way to go to the zoo. But it was just too dangerous.

One morning not long after the girls and Mohammed had left, he went for a stroll down the road, looking for diversions. It was a warm day, the baking heat of summer just passing. For something to do, Luay turned into the park that lay opposite the house, behind the mosque his father had built. The Americans had built the park in 2006 as they tried to restore order to Mosul after the invasion. It occupied roughly an acre, and was shaped like an egg. A perimeter of trees and bushes shielded a path on its perimeter. On the right side from the main gates lay a few carousels, which rarely worked and made an almighty racket when they did. In the centre was a pool with a hump-back bridge running over it, which in the summer was home to flocks of migratory birds and a few pedalos. Next to it was an enormous statue of a golden coffee pot. The left side of the park was empty but for a few trees and a shed.

The park was usually closed in the morning, but today – as Luay walked through the gates – it was open, and busy. A few Daeshis in their Kandahari outfits were walking around the asphalted area by the carousel. Through the trees, Luay could see a crowd gathering on the left side of the park. He sloped over to check what was happening.

'Luay,' shouted one of his friends. 'Come here. You've got to see this.'

He walked through the scattered crowd and saw three cages on the ground. Inside were a bear, a bear cub and three lions. One of the lions was small, not much more than a cub.

Instantly, Luay knew it was Zombie. He stood frozen still. Through some bizarre miracle, the lions and the bears from the zoo by the Tigris had been moved away from their home on the river to the park right in front of Abu Laith's house.

'They came on a lorry this morning,' his friend said. 'They unloaded them with a crane.'

Luay turned to run home. His friend shouted again. 'They're going to start a zoo here,' he said. 'They've brought the animals over. Apparently Daesh opened a training base in the place where they used to keep them, so they have to move.'

Luay, crowing with delight, ran home, swerving through the gates and into the house. 'Baba,' he shouted, crossing the courtyard into the living room. 'Zombie is in the park, and so are Mother and Father, and two bears.'

Abu Laith, who had been reclining on the sofa, jumped to his feet with an agility rarely found in larger men. He remembered later feeling a deep sense of satisfaction, and a strange lack of surprise, as if he had known the lion would come to him.

'Lions,' he shouted. 'I knew it. Zombie is back.' And without another word, he barrelled out of the living room door to see his lion.

Four days after the girls left Baghdad, they were driving across a stretch of desert much like the one they had been driving across for the last twenty-four hours. The inside of the car was hot and dusty, and their eyes and hair itched from the desert sand. Lubna, crying children in her lap, was about to snap. Driving through the endless plains of the Iraq–Syria borderlands, she envisioned the car as a tiny speck on a huge, empty cloth. If they broke down, they would die in the heat or be captured by the jihadis who would probably kill them. If they were captured, they might be killed. Their cover story was that they were on their way to a funeral.

Day and night, Lubna had scanned the horizon for attackers, who they might see coming from miles away but would never be able to escape. No one spoke much, and after a while the heat and the boredom dulled the fear.

Now, on the fourth day, they were still in the desert, but the villages had become plentiful and lanes became roads. They had gone past Isis territory proper into the no-man's-land between the government and the jihadis.

'There,' said Dalal, peering out the window. 'There's the army.'

Ahead, the siblings could just make out a line marked in the desert, which grew into a checkpoint surrounded by armoured vehicles. 'Put music on,' Dalal shouted at the driver, pushing her headscarf back. 'Open the windows. Drive slowly. We need to show them we're not Daeshis.'

She turned to the girls. 'Ululate,' she commanded, her sisters remembered later. 'Look happy.'

Lubna and Oula put their hands over their mouths and sang out the shrill, celebratory call Iraqi women use at weddings. They grinned through the windows. Ahead, they could see the soldiers at their position.

'Assalamu aleikum,' said Dalal, when they pulled up to the checkpoint, music pounding, her sisters ululating in the back. 'I work for military intelligence.'

13

Imad

1973

The dogs at the Polish house up on the edge of Zuhour were fenced in by necessity. Their barking spread fear around the neighbourhood so acute that few dared walk past them on the street. The few times snotty local boys had tried to go up to them on a dare the animals had launched straight for their throats, jaws snapping. As a result, the street by the Poles' house – a mansion with a garden near the office of the sulphur company they ran – was almost empty.

Imad had been doing his usual rounds when he heard about the wild dogs. He dropped by the house, looking for trouble. Barking met him as he walked up to the fence that ringed the compound. Intrigued, Imad marched up to the gate. The dogs, he saw, were German shepherds, a breed he recognized from pictures in books. He was not scared of them.

They were playing with a young woman, about twenty-five, who was wearing only a bra and shorts. She had blonde hair and was making the dogs stand up on their hind legs and hold her hands with their front paws. Imad barely glanced at her, though it was highly unusual for a woman to be running around a garden in her underwear in Mosul.

The young Polish woman watched as the boy strode up to the gate. He had a shock of red hair atop his thick-set face. He was short and stout, and looked very determined. Thinking he might be another European, she waved at him to come over.

'I'm here to meet your dogs,' he said, in Arabic.

'There's no point,' said the Christian Moslawi who served as a translator for the Poles, who was standing in the garden. 'They'll bark a lot if you come close.'

The boy ignored him. 'Hello,' he said, addressing the Polish woman. 'Can I come in and meet them?'

The Christian translated.

'Are you not scared of them?' asked the young woman, who took her dogs with her everywhere she went, and had grown used to how the locals would scatter when they came close.

'Not really,' said Imad, though he was in fact a bit scared. He walked through the gate into the garden and crouched low to the ground.

For a few minutes he sat at a respectful distance, whistling gently. The dogs quietened. Then, as slowly as before, Imad walked over to the dogs. It was as if he had spoken to them. All their anger gone, they licked Imad's hands and howled throatily with pleasure.

The translator was mystified. 'Why didn't they bite you?' he asked Imad.

'They couldn't bite me,' said Imad. 'They respect me, all the animals do. That's why they call me Abu Laith.'

The young woman understood. From then on, Imad came to the house almost every day. He helped her feed canned meat to the dogs, and play fetch with them. Soon, he could stand up with them on their hind legs, holding their front paws as the woman did.

It was his first encounter with an important truth – that good things come from being kind to animals.

14

Hakam

At first, no one believed the posters that went up around Mosul – pasted on street corners and on walls around the markets a few months after the militants had arrived. In large black letters on white paper, they specified that the smoking of cigarettes and narghile was banned throughout the Islamic State, and that any vehicles transporting tobacco would have their contents confiscated. Smokers could be subject to lashes or imprisonment.

The news was met with confusion. While the people of Mosul were, of course, technically aware that cigarettes were banned according to the Quran, few had ever thought of acting on it. While only a minority of people drank alcohol, smoking – cigarettes and narghile – was built into the fabric of daily life.

In the rickety streets of the Old City and the wide boulevards of the eastern side, cafes spilled on to the streets, the air filled with smoke from the great glass narghiles that waiters brought out, running back and forth with a basket of glowing hot coals to light them. In the evenings, the smell of petrol and fish from the Tigris mixed with the sharp mint or apple odours of the local tobacco blends. In the afternoons, men sat drinking sweet tea on street corners, puffing away at Winstons or Kents.

When the signs first went up no one took them seriously. It would take more than a few posters, Hakam's friends said, to wean Moslawis off their smoking habit.

At first, the hisbah would come up to smokers and give them well-meaning – if badly received – lectures about the dangers of smoking: the lung cancer and the emphysema. After a while, when everyone ignored them, they started taking packets of cigarettes off smokers.

But soon rumours spread of shop owners who sold cigarettes being hauled away, and not coming back. Smugglers, long a fixture in Mosul, started doing a booming trade in cigarettes, sending the price of a pack of twenty soaring from less than a pound to more than £13. Hakam rationed himself to one cigarette every three days in the garden, at night time, under the stars, hoping no one would smell it from outside the walls of the house.

The narghile cafes shut down, and so – more slowly – did the regular cafes. Some were still open, but customers were plagued by wandering hisbah, who would stop to measure beards – which had to stick out from the bottom of a closed fist – or trousers. Few women who had any other option left the house any more. Little by little, the life was drained out of Mosul.

The boredom had grown worse lately for Hakam, a strangling inertia permeating every day. With it came the anger, and the recklessness. In the last few months, the militants had grown more petty, justifying everything from the confiscation of shop mannequins to the lashing of smokers, all with mangled references to the Quran. Hakam's father, who knew the intricacies of sharia better than almost anyone in Mosul, would berate them whenever they presumed to challenge him.

It was even harder to stomach being lectured when it came from the westerners – some very recent converts – who had turned up in Mosul in the last few months. Hakam found it absurd that people from Norway or Zimbabwe had come to Mosul to find their extremist utopia.

When he bumped into them, as he sometimes did, he thought of excuses to give them directions or ask how their day was going. One afternoon, he had been in a video game store when a blond man walked in, greeting everyone in bad Arabic. He seemed only to know a few words, and everyone in the shop muttered in reply.

'Assalamu aleikum,' he said to the shop owner, before switching to English. 'Do you have *Call of Duty*?'

The terrified shop owner had no idea what he was saying, and looked around wild-eyed for support. Everyone else in the shop began to filter out, or became very interested in the games displays.

Hakam stepped in. 'Can I help?' he asked, in the impeccable English he'd learned from chatting to his foreign friends online.

The foreigner smiled and thanked him. A few minutes later he left, *Call of Duty* in hand, and everyone in the shop could breathe out.

That was the first encounter, but not the last. To date, he had seen Frenchmen, Americans, Brits and Indians, as well as men and women from a plethora of nations across the world. All wore the full garb of the Islamic State, and the men were armed. Compared to some of the locals, who had signed up for the power and the money, these international volunteers behaved like members of a cult – following their new rules far more closely than most of the Moslawis did.

Hakam could not get his head around the fact that they had colonized his city as their promised land.

15

Abu Laith

THE PARAPET THAT RAN AROUND THE EDGE OF ABU LAITH'S roof was about 3 feet high and 20 long – rosy pink cement shaped into a buttress that jutted over the side of the house. From there, anyone crouched on the roof with the crown of their head sticking over the top would be all but invisible to anyone working in the new zoo in the park outside his house.

It was mid-afternoon, and Abu Laith had been engaged in his usual roof-based zoo management. He had been lying low since the men had come to his house from the mosque, worried that they would arrest him if they saw him in the zoo. But his safety came with an enormous drawback – he couldn't spend time with Zombie, nor with Mother and Father, or Lula the bear and her son, whom the local children had named Warda.

In the week since the lions had arrived at the new location outside Abu Laith's house, no one competent had appeared to look after the animals properly. Someone, Abu Laith reflected furiously, had to keep an eye on things. Ahmed, the worker who had kept Zombie and his sibling in a pen in his garage, appeared to be in charge of the zoo following its enforced move. He didn't spend much time there, and he certainly, Abu Laith thought, didn't care much for the animals beyond collecting the 500 dinars that people paid to go and look at them. Other than him, there were a few workers who trimmed the verges, ran the carousels and occasionally gave the animals some food.

Technically, Ahmed was the eyes and ears on the ground of the zoo owner, Ibrahim – managing the zoo in his boss's absence. Ibrahim lived in Erbil, and had nothing to do with the day-to-day running of the amusement park.

Since the zoo had been moved to its new location, Abu Laith had watched animals arrive almost every day. Peacocks, monkeys – including a baboon – and an ostrich with long black plumes had been moved in on lorries, squawking and posturing with indignation. Then there had been the goats, and the fox, some guinea pigs and a squirrel. Snakes had come in glass boxes, while two Shetland ponies were left to run free in a small enclosure.

Most of the animals were arranged in cages around a flat cement area on the left of the zoo as he faced it from his house. He wished he could go and see them in person. Since a brief visit on the first day when the animals arrived, when he had managed to catch a joyful glimpse of Zombie, he hadn't been inside the zoo. On the edge of the park, between Abu Laith's house and the zoo, lay the mosque. As he hadn't started praying, like they'd told him to, he couldn't risk going past it and being spotted. Instead, he waited on the roof, spying on the goings-on inside the zoo, biding his time until he came up with a plan.

But he had almost been spotted several times by Ahmed and the Daeshis moving around inside the zoo. What he needed was someone on the ground to keep an eye on things for him. 'A spy,' thought Abu Laith, as he moved away from his post.

He went downstairs, and started to plot. He already had Luay, whom he could send inside the zoo. But eager as he was, the boy was still about as threatening as a cocker spaniel. He needed a lion – someone who could keep the animals safe, and keep a watchful eye on Ahmed. Given an inch of space, Abu Laith considered darkly, that man would sell the animals off to the highest bidder.

He knew what he had to do. Taking a chair, he sat himself by the front door like a man on a throne. The gate was half-closed, so he could observe happenings on the road without being seen himself. He sat, sipping his glass of tea, waiting for a suitable spy to appear – someone he could send into the zoo on his behalf.

It took until the sun was setting and the road was baking with the heat of the day. Luay had joined him, and they were watching the road together. A few young men had walked by, but Abu Laith had dismissed them all. Too short, too thin, too weedy. They needed someone with real heft, who could – in a pinch – stop any attempted thievery of the animals and supervise Zombie's dietary regime.

At around 6.30 p.m. a hoarse village voice could be heard yelling further down the road.

'That's Marwan,' said Luay. 'I know him from around here. He's a good guy, but he's pretty sketchy.'

Abu Laith looked interested.

'He drinks sometimes,' said Luay. 'And his dad is a real piece of work. But he's strong.'

Taking stock of the approaching figure, Abu Laith stood up and wandered over to the gate. The young man was about the same age as Luay, barely out of his teens, slim but solid. His forearms were smudged with home-drawn tattoos, mossy green reliefs on his skin. His complexion was burned dark as a coconut shell, and he looked furious.

He was, Abu Laith thought, perfect. 'Assalamu aleikum,' he said, sidling up to him.

The man glared at him, and nodded at Luay. 'Wa aleikum assalam,' he said.

'Where are you from?' Abu Laith asked.

'Hasan Sham,' he said. It was a mostly Kurdish village at the edge of Daesh territory. Though Marwan was Arab, he had learned to speak Kurdish growing up there.

'Do you have a job?' asked Abu Laith, knowing full well that he didn't.

'No,' said Marwan. 'Do you know of anything?'

Abu Laith, who had come up with a plan that pleased him greatly, gestured at a pile of bricks by the gate. 'Would you mind helping us move some of these?' he asked, keeping an eye on Marwan's reaction. Luay loitered nearby, unwilling to interfere in his father's work.

Marwan agreed. In a moment he was humping bricks between two piles near the gate. His arms, Abu Laith saw, bore up admirably, and he seemed to be able to follow orders.

'Good,' he said, as Marwan put down an armful of dusty bricks. 'Come inside. Welcome.'

They retired to the shade, and sat down together. Marwan was, he said, twenty years old, and had come here with his family to make some money in the city. He needed work very badly.

'I'm looking for a spy,' said Abu Laith, as they reclined amid the din of the sparrows that nested in the courtyard walls. 'I need someone to work in the zoo and make sure it's running properly.'

Marwan looked sulkier then ever. 'I'm not here because I like animals,' he said, as Abu Laith eyed him shrewdly. 'I'm here because I need work.'

Abu Laith considered him. At least he was honest. He could deal with a man who wanted money. If he was poor, he would treat the animals well, otherwise Abu Laith wouldn't pay him. It was the people who wanted power, or those wimps who would let themselves be bossed around by others, that he really couldn't trust. 'That'll change,' Abu Laith later remembered telling him. 'Now you listen to me.'

Marwan left that day with his head spinning. Zombie's diet regimen had taken at least an hour to explain. He didn't know that cleaning a peacock's cage could be so complicated.

But there he was, anyway, a newly inscribed apprentice zookeeper. The next day, he would offer his services to Ahmed, saying only that he lived in the neighbourhood and was looking for work. The mad old man had been very clear in his instructions. Ahmed was looking for cheap labour, and would no doubt hire the strong young man. But Marwan would report only to Abu Laith. He would disregard the feeding schedule laid out by Ahmed, who would pay him a few dollars a day to work in the zoo. Instead he would report back to Abu Laith on what the lions were being fed, and he would send him out with a new shopping list. Abu Laith's money would make up the difference, supplementing his meagre wages and buying meat for the animals.

The old man was, Marwan thought as he walked home, undoubtedly crazy. But he liked him, though he wasn't quite sure why. That day, for the first time, Mosul started to feel a bit like home.

Within a couple of days, Marwan had started work at the zoo. In the mornings, under the dappled shade of the trees, he swept the cages and, warily, fed the lions with meat that Abu Laith subsidized. But it wasn't enough – they were always hungry, and slightly sick. Most mornings Zombie's father would be sleeping – ribs jutting out a little through his skin. Zombie would be sunning himself in his cage, fur glinting a deep russet.

His mother would sit low on her haunches, staring at Marwan with a distinctly malevolent air. He did not, he often thought, as he filled up buckets of water from the tap, trust her one bit. As he worked in the yard sometimes, he fancied he could feel her eyes following him. Occasionally she would let out a growl that sent him scurrying. Lula the bear and her son Warda were much more polite. They were quiet and still, though Lula flinched at every sound. Abu Laith said that she had been afraid when she lived in the zoo by the Forest because of the sound of fireworks that were set off there so often. Her mate, who Abu Laith thought had protected her, had been left behind at the Forest, and now Lula was alone with her cub.

The monkeys spent much of their time screeching at Marwan, making him jump. Nusa, the oldest, had fur as red as Abu Laith's hair. Marwan, who wasn't particularly well mannered himself, had been pretty appalled at how badly she treated the visitors to the zoo. Whenever someone would walk in front of her cage, she would throw herself at the mesh and scream at them until they ran away. Marwan did nothing about it, unless the visitors were young women who he supposed might be pretty under their *khemmars*, in which case he would gallantly chase Nusa away to go and shriek in her cabin.

The baboon presented a different challenge. He was an oaf of a creature, lazing around his cage. But sometimes, if there were a lot of women around, he was spurred into action. Leaping up, he would spread his legs and wave his penis wildly in their direction. Cries, screams and chaos would ensue.

If things were getting a bit dull, Marwan would click his tongue at the baboon to get him going. When he told Abu Laith afterwards, at one of their regular debriefings, the older man would cry laughing. 'He's a ladies' man,' he would shout, rocking on his sofa.

There were a lot of things to tell Abu Laith. First there was the food. Ahmed and the other guards gave him bags of leftovers from who knew where – probably restaurants, since rice featured so heavily in the diet. At first, Marwan had tried to feed the rancid mix to all the animals. But none of them, except the two small goats and the fox, would eat it.

After consulting with Abu Laith, Marwan started spending each morning sitting on the floor by the lions' cages separating out pieces of meat from the rice, or plucking beans out of sauce to feed to the animals. The monkeys liked olives and pickles – any kind – while the guinea pigs liked lettuce. The lions, it transpired, did not like rice. With Abu Laith's money, Marwan would buy a whole sheep's carcass from the illegal slaughterhouse by the old city walls and feed it to the lions, who would strip it delicately with their teeth, holding the skinned body between their paws.

Most days, Marwan wouldn't even get paid by the zoo. Ahmed or one of his associates would give him a sandwich or two, with the implication that he should be happy to have it. Other days they would give him a couple of thousand dinars – about $2.

Without Abu Laith paying him, Marwan often thought, he would not be caught dead anywhere near this zoo. He had never cared about animals – barely even thought about them. But as he sat every day with the rancid food, picking out bits of gristle, he started to feel a bit depressed whenever he had to feed it to the peacocks, or the goats, or the fox – who would have probably preferred something more recently alive.

The two Shetland ponies, who scampered around in their paddock with their fringes bouncing, ate grass and hay. So did the goats. The peacocks and the pigeons had rice, which wasn't the right thing to feed them, but it was the all they had. The fox ate everything, but the snakes – long as a man's arm – were extremely picky, and only liked mice, which were hard to catch. None of the animals, except for the goats and the pigeons, liked old bread.

Following Abu Laith's instructions in their spirit, if often not their content, which was too complicated to remember, he washed down the cages with water every day, and tried to feed the bewildering array of animals something that would, at least, not make them sick.

Most days, after he had finished work, Marwan would go and report to Abu Laith. With each new piece of information, Abu Laith would get angrier and angrier until at some point he would explode in a deluge of curses and promises of extreme violence. Often, he would tell Marwan to do something so bizarre that he could barely fathom it. Once he had told him to buy Lula a toy because she was bored.

As Abu Laith struggled to think of ways to entertain the bear, he remained slightly oblivious to the more pressing difficulties facing his family. One extremely hot day, Lumia was walking down the street, too warm in her abaya and *khemmar*, feeling nauseous. Every so often she would lift up the black fabric over her mouth and take a gulp of hot air. But the diesel fumes were making her drowsy, and she was worried she might fall. As she walked, she later remembered, she was wishing as hard as she possibly could: 'Please let it be low blood pressure.'

That afternoon, she had been clanking about stirring the ever-boiling pots and cleaning the ever-filthy children as they groped around her feet, when she felt a horribly familiar feeling. It was a sort of grinding sickness in her belly that she knew very well. She had a headache, and sat down too tired even to panic much. As she sat, she counted. Then she put on her layers of black, pulled on her gloves, and walked out the door.

The clinic was a two-storey building about ten minutes' walk from the house, fifteen in the ferocious summer weather. Downstairs was a pharmacy that sold painkillers, antibiotics and erectile dysfunction pills. The doctor's surgery was on the first floor. Lumia had been there before, and walked up the stairs sweating in the heat.

By the door, near where the receptionist sat, stood a man in a Kandahari outfit and a beard holding an unusually long rifle. He looked Asian, and Lumia wondered what the hell he was doing there. Next to him was a woman with her son, who looked about ten and was wailing, clutching his stomach. The doctor, or at least someone who sounded like the doctor, was standing just inside the door, swathed in black.

'Please treat him,' the boy's mother begged. 'Please. He's so sick.'

'I can't,' said the doctor. 'I can't, they won't let me do it.'

The guard stood silent, ignoring the scene in front of him.

'He's just a little boy,' the mother said, as her child cried. 'Please treat him.'

'I'm sorry,' the doctor said, and she sounded contrite. 'He'll have to go to a male doctor. I can't touch him. They won't let me.'

The mother, who was now crying herself, picked up her son and carried him screaming down the stairs. Lumia stood aside for her, and went through the door into the office, shaken.

'I'm pregnant,' she told the doctor.

'Again?' the doctor asked, sounding exhausted. The last time Lumia had been here, she had been pregnant with Mo'men and swollen with excitement – the doctor professional and the clinic well-equipped. Now Lumia just wanted to lie on the floor and cry. But there wasn't time for that, so they did an ultrasound, which showed the baby where it lay, heart beating.

'I'm pregnant,' Lumia said to Abu Laith as she plodded through the front gate, pulling her *khemmar* off. She sounded furious, and not a little accusatory. 'I'm pregnant and I don't know what we're going to do.'

Abu Laith grinned. 'That's wonderful,' he cried. 'Why are you sad?'

'What the hell is wrong with you?' screeched Lumia. 'I can't have a baby now.'

Abu Laith, whose dreams stretched to little more than

being surrounded by a large horde of children and animals, felt the pride swell in his chest again, as it had for his last twelve children. 'Don't worry,' he said, jubilant. 'It'll be OK. We're having another baby.'

But Lumia, who felt very faint, could not be happy.

16

Imad

1978

By the time he was eighteen, Imad had discovered an immutable fact: animals were better than people.

There was one exception, and her name was Sara. She lived in Baghdad and she loved animals and she loved him. Since they were small, Sara and Imad had been promised to each other by their families, who were part of the same extended clan. From when she was about fifteen, and Imad a couple of years older, he started making excuses to come to Baghdad to visit her family. He would walk her to school, wearing his flares and tight, pointy collared shirts, and pick her up after her classes.

They had been seeing each other in secret for over a year when long-running tensions between Iraq and its neighbour Iran boiled over and Saddam Hussein launched an invasion across the border into southern Iran in 1980. Conscription was declared, and every day young men were dying, and cities were being bombed. Food prices rose, and people became desperate.

In the midst of it, Sara's father – her family's breadwinner – died. Her mother, who didn't know what else to do, told Sara she would be married off to an older relative on her side of the family. Sara would raise a dowry that would keep the family stable, and she would be safe from the attention of other men.

Horrified, Sara flung herself on Imad the next time he came to Baghdad. 'We have to get married now,' she had sobbed. 'You have to force your parents to allow it, or they'll give me away to someone else.'

Imad told her that everything would be fine. They were young and in love, and it seemed simple. Imad would go back to Mosul and send his parents to Baghdad to ask Sara's family for her hand in marriage. They would all come back to Mosul together, an imam would read from the Quran and they would sign a marriage certificate. They were promised to each other. There shouldn't be a problem.

But Imad didn't know that the older Sara became, the more his parents disapproved of her. She was flighty, and far too interested in insects and cats and all the strange things their son loved. Their son needed a solid Moslawi woman to ground him, they thought.

His mother decided to play a trick on Imad to keep him at home. One morning, after Imad had badgered them for weeks, they left to go to Baghdad and ask for Sara's hand. But instead of going to her parent's house, they stopped off at the home of some other relatives, slept there a night, and drove back to Mosul without speaking to her. Sara had been waiting for them to come, but they never did.

Back in the house in Mosul, Imad waited in the living room, which had been lain with plates of kibbeh and rice, for the engagement feast to start. His parents came in the door, looking glum after their trip to Baghdad. There was no one with them.

Imad's father was deeply downcast. 'Sorry, son,' he said. 'She said no.'

Imad didn't believe him. He told him he was a liar.

'She said no,' his mother insisted. 'She said, is Baghdad so empty of men that I would marry Imad?'

And the young man felt everything splinter around him.

For three days, he locked himself in his room and didn't eat or sleep. When, finally, he got out of bed, Imad wasn't the same. Sara's betrayal had driven him further away from the company of humans, whom he thought untrustworthy and cruel-minded.

The pain had been lessened slightly when he sent word to Sara's relatives that he would kill them if they ever set foot in Mosul. Such threats were relatively common, and assiduously carried out. The clan in Baghdad – though they were none the wiser about the reasons behind the threat – thought it best to stay away from Mosul.

At first, Imad had refused to believe that Sara had anything to do with the rejection. But soon his mother provided the answers he craved. She was flighty, she said, and had probably fallen in love with someone else. It didn't take long before he began to believe that she had never loved him.

He needed a distraction, and one came very quickly in the form of conscription into the army to fight the Iranians. Within weeks, he was on a blustery hill-top near the Iranian border. He was hungry, cold and lonely.

Dreams of Sara and a quiet life with their animals were left behind. At night, when he lay in his tent in the mountain encampment, the thuds of mortar shells falling around him, he thought of his friends back in Mosul, and felt like he really needed a drink.

Imad was not interested in war. It all, he reflected, seemed a bit pointless, and a lot of work. More than anything, he didn't like being told what to do. He was sulky, avoided most tasks, and was quickly relegated to the mechanic's workshop, with its smell of hot iron and cigarettes.

Occasionally he would get roaring drunk on smuggled whisky and talk to the stray dogs that roamed the mountains. Though army rations were always tight, he saved some of his food to give to the dogs, luring them with rancid pieces of meat held out in his hand. Sometimes, he would feed them whisky, which seemed to make them pleasantly dizzy.

'Imad, why do you keep feeding the dogs?' his fellow soldiers would ask. 'People are hungry, why don't you feed them?'

Imad had only laughed. At the mess, no one would sit with him – the filthy Moslawi who ate with dogs. But Imad didn't care.

A few months in, he was bored of the knifing cold and decided to leave. He pulled a few strings with a commander, and he was shipped home for a spell without being officially signed out of active service. He arrived home to discover that his mother had found him a wife.

'Absolutely not,' said Imad.

But his mother didn't listen. The longer they waited, she knew, the more likely it was that Imad would go down to Baghdad to find out what had really happened to Sara. 'You'll like her,' she said, and went off to make the preparations.

As he wandered around the alleys of Zuhour waiting to be sent back to the front, Imad conceded that it might be quite nice to have a wife after all. He soon found out that she was called Muna. She was a few years younger than him and skinny and shy as a cat. Imad had never even heard of her before. He met her for the first time at their wedding. Muna wore a flowery red robe, and her hair was braided down her back. They cut the cake together, and Imad felt sick with the wrongness of it all.

The wedding rushed past in a blur of relatives drinking orange squash and the couple avoiding each other's eyes across a room divided by gender – the women ululating behind a curtain in their best dresses, the men smoking narghile.

By the time the guests were filing out Imad had disgraced himself by enquiring of the family – if it was all the same to them – whether he could marry Muna's twin sister instead. Muna was skinny, while her sister had cheerful, chipmunk cheeks and was round as a melon, which Imad liked. But she was already married, and Imad and Muna were bustled out of the hall before anything else could go wrong, her relatives reassuring Imad that she would get fatter as soon as she had children.

'With twins, it's like an apple,' one of her family had said. 'When you cut it in two, each half still tastes the same.'

Twenty-three days later, the couple moved into their new home on a quiet street in Zuhour. Inside was a courtyard spread out onto airy, newly painted rooms, and in a coop under the stairs were a flock of pigeons looking down disdainfully on the rabbits, dogs and chickens that scuttled and pecked their way around the courtyard.

'I'll look after them,' Muna had promised, and set to work.

So they settled into their new life – Imad, between his months-long stints being sent back to the front line, fixed cars and bred pigeons. He started buying chicks in the Ghazal market in Baghdad for 25 dinars apiece. When they had grown, he would sell them for 400 dinars, but not until after they had borne thirty chicks each.

The market was a flapping, screaming primal hole of a place. On Fridays, the streets ran with blood and gore and all the effluents of a thousand life-forms, growing and dying and making an unholy racket. Imad loved it. When the pigeon chicks were safely stowed in the driver's seat, chirping in their boxes, he would drive back to Mosul, wind streaming through the windows, feeling extremely pleased with himself.

But then the call from the army would come again, and he would go back to the front, leaving Muna to look after the animals.

17

Abu Laith

Abu Dalal was a devout man with a kind face, who didn't want any trouble. He was an imam's assistant who lived on the same road as Abu Laith. When Daesh had come, he had kept working at the mosque, though he didn't agree with what the newcomers were saying.

One day, however, as he went to the mosque for afternoon prayers, Abu Dalal found himself in the unusual position of seeing something so terrible that he waded right into trouble of his own accord.

It was 4 p.m. about six months after Daesh had arrived when he left the mosque and knocked on Abu Laith's door, hoping no one would see him.

'What can I do for you?' asked Abu Laith, after they had sat down in the living room.

'I've got to tell you something,' said Abu Dalal.

The usual crowd had been inside the mosque – a mix of Daeshis and local people Abu Dalal had known for years. Together, they rolled their prayer mats out and faced Mecca, each beginning to pray. As they finished, Abu Hareth – the pale, large Daeshi whom Abu Laith had yelled at in the mosque – had called everyone to attention. He had been telling everyone to report their neighbours if they didn't go to the mosque, and to take down all the pictures in their houses.

'Do any of you know Abu Laith, who lives across from here?' Abu Hareth had asked. 'He is an infidel. He drinks alcohol, and he never goes to the mosque. We've warned him before, but he hasn't listened. He isn't at the mosque today, and hasn't been on any of the other days.'

The worshippers had listened.

'Just so you all know what happens to infidels, I'm telling you now,' Abu Hareth said. 'I'm going to slit Abu Laith's throat, and sacrifice him like a sheep at Eid.'

Lumia sat, terrified, as she listened to Abu Dalal's story. Abu Laith, however, was furious. 'A what?' he shouted. 'A sheep?'

Abu Laith had known that some of his neighbours didn't like him. They would sniff when they saw him coming back to the house carrying clinking bags from Bashiqa, though he made sure to never drink or be drunk in public. More than anything, they resented having been woken up in the middle of the night for decades by his brood of cooing, barking, screaming animals.

But this was too much. He had built that mosque with his own money, and now the people who prayed there had shopped him to Isis. For weeks, Abu Laith had stayed out of the street, hoping that the Daeshis at the mosque would forget about him, and assume he had run away. Someone must have told Abu Hareth that he, Abu Laith, was still at home.

'You have to leave now, or they'll come and find you,' said Lumia, who had listened in disbelief.

Abu Laith pondered his options as his family panicked around him. He would have to go into hiding. His cousins had a farmhouse on the outskirts of Mosul where he could go to ground until the immediate crisis blew over. But as he sat there, inspiration came.

When a gazelle is being hunted, the worst thing it can do is split off from the pack. Survivors keep low, and blend in. 'I'm staying,' he announced to Lumia. 'They'll never think to look for me here.'

She tried to protest, but he jumped up, energized by his audacity, and ran upstairs to investigate the attic.

The house had four floors, but the family – for no particular reason – inhabited only the bottom two. The third was empty except for a broken cabinet that stood in the dust by a window. The staircase that led up from it was built of uncovered breeze blocks. The fourth, the attic, was little more than a conduit to the roof – the big, flat space where Abu Laith spied on the zoo and bred pigeons.

No one, Abu Laith reasoned, would come to the top floor, which was a single room, if they saw that the one below it was abandoned. He unlocked the door from the attic on to the roof and left it open. That, he thought, would be his escape route, if it came to it. He could bolt over the side of the house, on to next door's roof, and down into the street. They would never, he thought, find him there. He would hide in plain sight, like a lion camouflaged among the long grasses.

But first he would have to warn the more reliable and friendly neighbours. Abu Laith went downstairs and looked out of the front gate. The street was empty. He ran as fast as he could, circling back until he came to the house directly behind his own. The people who lived there were relatives of his first wife, and agreed to help him. If he jumped on to their roof from his own, they wouldn't give him up to the Daeshis.

After a round of hand-shaking, Abu Laith bolted back around the corner to the house next door to his. The head of the family who lived there had been a general in the army, but died a few years ago. He had been a member of the Shia Shabak minority, like Abu Laith's drinking buddy Sheikh Hassan Beg. Isis would certainly have killed him. At home lived his three sons and their mother, who was Sunni. They knew Abu Laith well – after their father had died, he had helped them, and brought them watermelons in the summer. They, too, would help him if he had to flee his own house.

Good deeds paid off, thought Abu Laith, feeling inordinately pleased with himself as he snuck back home.

The day wore on, and Abu Laith settled into the living room for the evening. Lumia was clattering around, furious at her husband for annoying the Daeshis so much that they wanted to kill him. But she was glad that he was staying. There wasn't much chance of them surviving without him.

Someone rapped on the door to the courtyard. The children, who were used to letting in anyone who called, rushed towards the door.

'Stop,' Lumia yelled, running to grab them before they tore the gate open. Luay ran out of the house, white as an egg.

'Is he upstairs?' Lumia hissed. 'Is he hidden?'

Luay saw Abu Laith, who had heard the knocking, lumbering up the stairs.

Outside, Lumia shooed the last of the children into the house and latched the door behind them. She grabbed a scarf from her bedroom, and wound it around her head.

'Who is it?' she shouted. The banging on the gate had intensified.

'We're from the *dawla*,' said a Moslawi voice. 'Open the door.'

Lumia stood by the gate, trying to calm down. They wouldn't come in, she told herself, if they thought she was alone. Even for arrests, they wouldn't allow a man to enter a house where a woman – a potentially uncovered one – was by herself.

'There are no men here,' she shouted. 'It's just me with the children.'

'Open the door,' growled a voice on the other side of the gate. 'Now.'

Lumia opened the door a crack, holding it nearly closed so all they could see were her eyes.

'I'm alone,' she said. 'You can't come in. My husband is in Qayyara. I don't know anything about him. I'm only his second wife, he has another house.'

There were five men outside the gate, all bearded, dressed in long shirts, and armed. 'Assalamu aleikum,' said one of them. 'We're looking for your husband.'

Lumia pulled her scarf over her face and tried to hide herself behind the door, so that they wouldn't see she wasn't wearing an abaya.

'We're coming in to check,' said one of them, a Moslawi, and pushed at the gate. His voice was soft, but his eyes looked furious. Lumia thought this must be Abu Hareth, the fighter who wanted to bleed her husband out like a sheep.

'I'm alone in here,' said Lumia, holding the gate at a glint. 'Are you going to come in to a house where there's a lone mother?'

The Moslawi hesitated.

'I just had a baby,' Lumia cried, in a moment of inspiration. The fighters would, she knew, consider her condition impure.

For a moment, she stared right at Abu Hareth, who glared at her. 'Let us know if he comes back,' he said. 'We'll come and check soon, anyway.'

They withdrew. Lumia closed the door and listened. She locked the gate, and sat down on the floor. A tide broke inside her, and she started sobbing.

'It's OK,' said Luay, as he ran over to her. The children had been hiding in the main bedroom, pledged to silence.

Lumia was hyperventilating, and it took her a moment to get her breath under control. 'Where is he? Did he jump?'

Then there was the sound of feet hammering down the stairs, and Abu Laith appeared, looking very pleased with himself. 'I'm OK,' he beamed. 'Did you miss me?'

18

Abu Laith

THE FALAFEL STALL STOOD IN THE CENTRE OF THE ZOO, next to the cafe and the ice-cream shop that was open all year round, except when it rained. It sold sandwiches filled with golden brown balls of chickpeas and herbs – deep fried in a great skillet of oil – wrapped in flatbread with garlic sauce, pickles and tomatoes, wrapped in greaseproof paper, to be eaten at one of the tables outside the cafe, which also sold egg sandwiches. Families would sit there, children picking at their food or drinking sherbet, the women lifting their *khemmar* a few inches off their face to sneak a bite. Next to them, in the pond, the pelican occasionally squawked.

Sometimes, Marwan had seen the women give up and walk over to the trees over to the right, check there was no one watching, and flip their face coverings up for a few moments so they could eat their falafel in peace.

Today, he was standing in his usual place by the animal enclosure, looking over to the fairground about 100 feet away on the other side of the park. Friday nights were the busiest night of the week at the zoo. While the rest of Mosul lay in darkness, the walkways and rides were illuminated by the flashing white and red lights of the carousels and the spotlights outside the cafe.

The first customers had started to come around 6 p.m., and by 9 p.m. the zoo was usually heaving. It was the only place for miles around where Moslawis could go to relax – the rides and cafes by the Forest were too close to the Daesh training camp, and – of course – they could no longer go to the Kurdish cities of Erbil to the east or to Duhok in the north.

So they came, the women swathed in black, the men with beards, the children bouncing from sugar and excitement. They picnicked, rode the rides, looked at the large gold statue of a coffee pot that stood in the centre of the zoo. The rides didn't play music any longer, because it was forbidden under the militants' laws, but the speakers were in regular use. 'Lost child,' a voice would clang across the park. 'Lost child over by the falafel stand.'

There was always a group of crying children there who had lost their mothers in the crowd, unable to tell the difference between them in their buttery-soft black cloaks, *khemmar* and gloves. After a while, an angry parent would come and pick them up, taking them back to the rides.

The speakers were also used to call people to prayer, when everyone had to leave the park to go to the mosque. There were few excuses for missing prayers, and Marwan met none of them, so whenever the call went out, he hid in a shed over by the animal enclosure. They had caught him once, but he had convinced them he was fixing something and hadn't heard the call, so they let him go with dire warnings.

From their accents, Marwan could tell that most of the families who came to the zoo were Moslawi – civilians out to have a nice evening. But he could never relax, or be sure.

Dotted among them were Isis families – sometimes locals or, more rarely, foreigners who spoke their strange, stilted formal Arabic. Once, Marwan had met an Iranian Kurd, and they'd spoken Kurdish to each other. The man had even been wearing traditional Kurdish clothes. It had made Marwan feel very strange.

Though he never spoke to the women apart from when he sold them their entry tickets, he could tell that many of the Daesh wives were foreign, too. Their kids looked like they came from every country in the world – skin that ranged from bone-white to ebony, yellow hair and Afros. From a few words, 'yes', 'no', 'thank you' he thought they were speaking English, and sometimes Russian and French, though he didn't know any words in these languages.

Sometimes, a big group of them would come on a day out. Once, fifteen Pakistanis, men, women and children, had come for a picnic in the sun, and to marvel at the lions. The men had been armed, and the women as quiet as they always were.

The young Daeshi children were just like any of the locals – completely unaware of how dangerous the animals were. Occasionally, one of them would launch a small body at the lion cage. Only Marwan's lightning reflexes, honed in years of street brawls, saved them from becoming lunch.

But the older ones had learned violence from their parents, and seemed to hate everything alive. They picked up sticks to prod the lions, and threw stones at the bears. Nusa the monkey was traumatized from the beatings the children gave her, and screeched more than ever. There was nothing Marwan could do, he knew, other than stand there, looking big and impotently angry.

Abu Laith, when he had told him, had been so furious that

Marwan had almost bolted out of the house. 'Who gives a shit if they're Daeshis?' he had shouted. 'You should break their hands, all of them.'

Marwan, who had a more finely honed instinct for self-preservation than his benefactor, ignored his advice and carried on standing by as the Daeshi children harassed the animals. At first he hadn't cared that much. But day by day, it hurt a little more when Zombie yowled or Lula, terrified, pushed her back up against the bars. He was going soft on the animals, he realized with some surprise.

The families were not the only visitors to the zoo. When it was at its busiest, a Toyota pick-up would pull up at the gates, carrying half a dozen Isis fighters armed with rifles. They would split up into pairs and patrol the area, looking for exposed ankles or single men – who weren't allowed in the zoo. Most of the Daeshis had long beards and wore long Kandahari shirts.

One evening, as Marwan loitered by the animals' enclosure, he watched Zombie, who was sitting in his cage, and Lula the bear, who often spent her time trying to roll on the floor of her own cage, without much success. Warda, her cub, mostly just scampered about, getting in the way and yelping whenever his mother rolled onto him.

With a thrill of panic, Marwan remembered that he was supposed to have dug Lula a pool to wash in. Abu Laith had made him promise to do it a few days ago, and had said he wouldn't get paid if he failed. Abu Laith had seen on the National Geographic channel how bears needed to roll around in dust and water, and decided Lula and Warda urgently needed a pool.

Marwan needed to get paid. There were smuggled cigarettes

and sandwiches to buy, and money to give to his family. Despite the urgency, Marwan had forgotten about the pool – between cleaning the cages and trying to flirt with girls it had dropped off his list of things to do. It was a big undertaking, and probably pointless. He couldn't let the bear just run free and go swimming. The problem was that he had promised to go to Abu Laith's house after work that evening. He would just lie, he thought, and do it another day. The old man wouldn't know the difference.

He had, he explained later, been starting to rely on Abu Laith. His own father just ignored him. But Abu Laith gave him the time of day, even when he was shouting at him for making mistakes. He never seemed to write him off – even when he had been caught sneaking packets of cigarettes on Abu Laith's account at the shop, after begging the shop owner to write them down as cans of Pepsi or chocolate. Abu Laith hadn't been angry, just sad. He hadn't done it since.

Despite the fact that Abu Laith was wanted by Daesh, Marwan had never considered betraying him. He knew that while he and the others did this for money, Abu Laith did it because he loved the animals. Twice, now, Abu Hareth had stopped Marwan on the street, pulling up alongside him in his Mercedes to ask if he had seen Abu Laith. Both times he had said no.

It was almost midnight by the time Marwan finished up at the zoo and went over to Abu Laith's house. The old man was lying on his sofa, as usual. 'Is the zoo clean?' he asked, as Marwan walked through the door.

'Yes,' Marwan said, thinking of the unswept cages.

'And you dug Lula her pool?' Abu Laith asked, with a hard note in his voice.

'Yes,' Marwan said, trying to look nonchalant.

'Did she swim in the pool?' Abu Laith asked.

'Yes,' said Marwan, firmly. 'She really liked swimming in the pool.'

Abu Laith considered Marwan for a moment. 'Are you sure?' he later remembered asking.

'Yes,' whispered Marwan.

'Let's go and check, then,' Abu Laith shouted, cheerfully dragging Marwan with him out the door.

'You can't go outside,' Marwan said, as they made for the gate. 'They'll see you.'

But Abu Laith already had his head out of the door, sniffing the night air. The street was quiet, and the mosque where the Daeshis prayed looked closed. 'There's no one inside the zoo?'

'No,' said Marwan, who was feeling extremely nervous. The old man was going to get him killed by Daesh. If not, he'd probably murder him with his own hands after he found out he hadn't dug Lula's pool.

The pair paused by the gate. Then, moving stealthily, Abu Laith ran across the road to the gates of the zoo, a reluctant Marwan behind him. Inside, the old man checked that no one was following them. The night was quiet and warm. Abu Laith looked around the park with the air of a man who has just walked into his house to find it full of rodents and squatters. 'It is a disaster,' he exclaimed.

Quite what was a disaster remained unclear to Marwan as he followed Abu Laith through the zoo. Every few steps the old man would stop, exclaim at something, and keep walking. He had a habit of occasionally kicking a particularly offensive object – an amateurishly installed fence, or a badly planted bush – or making loud estimates of how much he assumed Ahmed had overpaid for them.

Marwan seriously considered running for it as they approached the animal enclosure.

'Zombie,' Abu Laith whispered. The lion was awake, as was the bear Lula. Abu Laith, as always, thought they looked extremely happy to see him.

For a moment, man and lion communed through the bars of the cage.

Then the quiet night was broken by a shout. 'Dirty,' yelled Abu Laith. 'This cage is dirty.'

Marwan froze. Abu Laith advanced towards him, furious. 'You are a liar. Look at this place. It's filthy. And how much have I been paying you?'

Marwan fumbled for an excuse, but nothing came out.

'And where is the pool?' Abu Laith yelled. 'You liar, you told me you had built a pool.'

Marwan later remembered just standing there, completely silent.

Across the road at Abu Laith's house, Lumia was settling into bed. Her belly had been growing at a remarkable rate. They had decided to call the baby Shuja, which meant brave – like the heart of a lion. Laith (the oldest) was already named after her husband's favourite predators. He wouldn't call any of his children Assad, which also meant lion, because he loathed the former Syrian president Hafez al-Assad.

Abu Laith was relentlessly optimistic, but Lumia was worried. Unlike with her previous children, she didn't know how she was going to feed the new arrival. Her husband couldn't work, could hardly leave the house, and they had barely any savings. The price of food had shot up because all the roads to Erbil and Baghdad were closed.

Then there was the more immediate problem of how she was going to give birth at all. 'I don't want a Daesh doctor,' she had told Abu Laith. 'They're all mad.'

All of Lumia's births had been complicated, and she couldn't give birth at home, without blood or midwives, though she very much wanted to. The hospitals, she knew, were still functioning. Though many of the doctors had fled, the majority of the staff at the hospitals were medical personnel who stayed in the city when Daesh came. But there were Daeshi doctors among them too. Despite this, Lumia could give birth at the al-Khansa hospital, as she had with her last five children.

Abu Laith, ignoring her protests, insisted on going with her when the time came. 'It's final,' he had said. 'It's not safe for you to have the baby at home.'

They would bring Ridha with them, too; Abu Laith's fourth daughter was a cheerful housewife who lived closer to the city centre.

'But what if they arrest you?' Lumia had asked. 'Then what will happen to the children?'

Abu Laith, as he was prone to do at these times, ignored everything Lumia was saying. 'I'm coming,' he said. 'And that's that.'

19

Lumia

The kitchen was in a state of crisis as the rain pounded down outside. Ownerless shoes flew through the air. Crawling toddlers, taking advantage of the breakdown in order, started eating bits of stray leftover food. Abu Laith, slightly dazed, was trying to take charge by shouting louder than all the others. For once, it wasn't working.

Among them, Lumia scuttled around, deathly calm, her contractions heaving. 'Luay,' she shouted. 'I'm going to have a baby. Grab the little ones. They're all dirty. They need a wash. There's a bag of rice over there, eat that while I'm gone.'

She finished wrapping the sandwiches she had made. It was nearly midnight, but none of the kids had gone to bed, and she couldn't make them. She turned to Abu Laith, who had disguised himself as a villager – scarf covering his red hair, loose Kurdish trousers, big, dirty coat on over a t-shirt – in preparation for their outing. 'Get the car,' she said, firmly. 'I'm going to have a baby now.'

Abu Laith was, momentarily, terrified. Lumia didn't think she had ever seen him so hesitant before. She thought of the way he had almost shouted at the Daeshis at the picnic eight months or so before.

'And you,' she said. 'You'll keep your big mouth shut at the hospital, or you'll get us into trouble.'

Lumia strode over to the bedroom, shouting orders over a wave of plaintive requests. Puffing to ease the contractions that were growing in her belly, she threw some clothes into a bag. She pulled on a headband over the front of her hair, so it wouldn't peek under her hijab. She put on a black abaya over her housedress, and tried to do it up at the front. It had been months since she last left the house, and fear was gnawing away at her. She didn't know what she would find in the city outside. Her hairline was sweaty as she tied on a black hijab, and tucked the edges underneath the neck of her abaya.

She tied the *khemmar* at at the back of her head. Another dig in the wardrobe, and she found her elbow-length gloves and some long black socks. Last of all, she took the piece of the material that hung from the back of the khemmar and pulled it over her eyes. She was hot, and everything looked dark through the scarf. She was ready.

'You're going to have a brother,' she remembered crying out over the tumult. 'And you all need to behave until I get back with him. Luay is in charge, and none of you are to fight. And if you don't eat proper meals I'll know when I'm back, and I'll deal with you then.' She was still shouting instructions when Abu Laith bundled her, already soaked from sweat and the rain, into the back of the car five minutes later.

'Don't worry,' said Abu Laith in a considered attempt of joviality, as they drove up to the junction at the top of the road. 'Just stay quiet. If they ask you anything, then just don't answer.'

She tried to breathe calmly as she looked out of the window. Everything seemed greyer than it had before; dustier and poorer. Her husband would be captured and probably killed if they were stopped at a checkpoint. She would be arrested, she thought. But it was all irrelevant compared to the jabbing, stretching pains that told her that if they did not hurry up, she would be giving birth in the street, no matter who tried to stop her.

They stopped outside Ridha's house. Abu Laith ran out of the car and banged on the gate. 'Come out,' he shouted. 'We need you now. She's about to have a baby.' A moment later, the gate clanged open, and Ridha – also swathed in black – shuffled out.

As they turned on to the main road west towards the hospital, Abu Laith slowed the car to a crawl. The streets were dark and wet in the autumn rain, empty but for the occasional men in their thobes and long beards. Most were normal Moslawis, who might usually have worn suits or skinny jeans, but now dressed according to Daeshi rules.

Even though they were out of Abu Hareth's domain, Abu Laith drove slowly. Daesh had, he knew, become tetchy with fear of infiltrators and spies, and would shoot if cars drove by too fast. Usually, they waved people through checkpoints without much trouble, as long as they were properly dressed. But sometimes they asked for identity cards, checking them off against a list of wanted people. His name was certainly on there. If they took him, Ridha and Lumia would be left alone.

Lumia – whose contractions were ratcheting up fast – screamed through her teeth.

Abu Laith put his foot down. They were just a few minutes from the hospital – a large structure of metal and glass where many of his children had been born. Then, it had been the best hospital in the city, staffed with some of the country's most highly skilled doctors and their expensive equipment. He hoped that was still the case.

In the back, Lumia was trying to breathe slowly, through her nose, as she had been taught by the nurses when her first child was born. Ridha, who despite the danger was happy to get out of the house for a while, was trying to keep her spirits up. 'It'll be easy,' Lumia remembered Ridha saying, cheerfully, as they swerved past the double-parked cars and fruit stalls.

Lumia only glared at her through the black cloth. This was her seventh baby, and she knew all too well how 'easy' it might be. One of her first had almost killed her from blood loss. 'We need to hurry,' she said to Abu Laith, who was raking the side of the road for the hospital entrance. He saw the emergency room, and swerved towards it. They stopped just short of the hospital doors.

Legs twisting into her abaya, hijab pulling back as the mess of black material wound itself into one awkward piece, Lumia pushed open the truck door and began to climb out. Abu Laith tried to help. She shrugged him off. 'Get back in the car,' she said. 'Stay here. Don't leave.'

Abu Laith looked over to the hospital door where two Daeshis were standing, dressed in thobes and tac vests, carrying AK47s. In Mosul, men were not allowed inside delivery rooms. Birth was something that took place outside the realms of men. Abu Laith, however, thought differently. Last time Lumia had given birth he had slipped the doctor some cash and gone in there with her, hallooing as another son was born. This time, he thought, there would be no chance. But he would try anyway.

Before he had time to mull things over, Lumia was already walking past the guard on the door, who made a vague attempt to try to stop her. 'I'm having a baby right now,' she hissed, and he recoiled.

Abu Laith followed, but the guard stopped him at the door. 'That's far enough,' he said. With a last, worried glance at his wife, Abu Laith went back to the car.

The two women went into the reception area, where a scattering of people were waiting to be seen. 'I'm having a baby right now,' repeated Lumia, much more loudly. She was paying even less regard to social niceties than she normally did. The Daeshis standing by the reception desk gawped, looking out of place in their desert garb in the clinical modernity of the hospital.

A male receptionist was behind the desk. He pointed. 'The maternity ward is that way,' he said.

Ridha helped Lumia along the hall. Up and down the corridor, doctors and nurses were rushing past them – the men dressed in white lab coats on top of their scrubs or suits, the women swamped in long abayas, running to surgeries or consultations.

Lumia pushed through into the maternity ward with Ridha on her tail. On the other side she stopped short, horrified. There, inside what was supposed to be the women's sanctum, was a middle-aged Daeshi with a beard and a gun. 'Stop,' he said. 'Where are you going?'

'Where the hell do you think we're going?' Lumia gasped. Her panic had gone, replaced by iron-clad determination to destroy anything that stood between her and a safe place to have her baby. 'I'm about to give birth.'

A female doctor, face covered and wearing a lab coat, ran

up to them. 'This way,' she said, spiriting Lumia towards a delivery room.

Lumia started to hyperventilate. The black chiffon of the *khemmar* felt as if it was choking her. She pulled it back, and the cold air reached her face. 'Put that back on,' commanded the Daeshi, as Lumia staggered towards the room 'There are men in here. Cover yourself.'

The Daeshi followed them until they ran into the delivery room. They could still hear him stomping outside, seemingly furious at being shut out of the ward he controlled. 'You can't close this door,' he said, in a southern Moslawi accent. 'I need to be able to see all parts of the ward.'

'I've told you,' the female doctor said through the door. 'My patients need privacy. Get away.'

Lumia sank on to a bench among some other women, who shifted out of the way. Her contractions were seizing her belly every few minutes, and she was struggling to breathe. 'It's soon,' she told the doctor, panting. 'But I'm worried about my husband, he's waiting outside.'

'Fine,' the doctor said. 'I'll examine you in a moment. And don't worry about him, worry about yourself.'

One of the other women on the bench, black-clad and uncomfortable, piped up. 'Why does she get to go first?' she said. 'We were here before her.'

'She's further gone,' said the doctor, bustling off.

The other women turned to Lumia. Only one of them had her face showing. She was very pale and beautiful, and her hair peeked out from under her hijab.

'My husband he love me,' she said, in broken Arabic with a thick Russian accent, continuing a monologue that had clearly been ongoing before Lumia and Ridha arrived. 'He in Anbar.

I so in love him. He love me more than any people love any people.'

None of the other women said anything. This woman was a foreign Daeshi, a muhajira, and she was dangerous. 'I become pregnant in Iraq, I follow him in Iraq,' she said, looking satisfied with herself.

'And did he take another wife, or a slave, or a Yazidi?' asked one woman, sneeringly.

The Russian just smiled. 'No,' she said. 'He loves me only. My husband big emir. His friend they selling buying slave but my husband love me only.'

The doctor came back into the room and gave the muhajira a glass of water, asking if she needed anything. Lumia thought she was going to be sick. She was about to give birth, but she'd be stuck in the queue behind this moonfaced idiot because the doctors were so scared of her.

'Why are you here?' another woman asked the Russian.

'I want be martyr,' she said, smiling. 'I–'

'I need to give birth,' said Lumia, who'd had enough of this. 'Someone help me.' The doctor ran over, and helped her to her feet.

Ridha turned to the Russian. 'Don't you understand we're all scared of you?' she snapped. 'Your husband is a jihadi, it's dangerous for us.'

Lumia nudged Ridha in the side, scared of what else she might say. 'Shut up,' she whispered. 'Come on. Let's get this over with.'

Together, they went with the doctor into a curtained room. Lumia had barely made it on to the gurney when she screamed. She was half-mad with pain. Someone gave her an oxygen mask, and she put it on over her *khemmar*, barely able to breathe. She would have to wear this thing even if she was dying, she thought. Amid the contractions, she noticed the room had suddenly gone dark. The electricity was out. The Daeshi with the gun was probably still at the door. She was certain she was about to die, with her child.

Outside the hospital, Abu Laith was sitting in his car feeling uncharacteristically tense. Lumia had been inside for hours now, and he hadn't heard any news since the last time Ridha had run outside to update him. She had told him Lumia said that he should go home – but he had ignored her and stayed. It would be pointless, he thought again, to ask the Daeshis standing outside the hospital what had happened to Lumia. They would only tell him to piss off, or ask him for his ID card. It was better to wait. Lumia knew what she was doing.

He settled comfortably in his seat, and stared out of the window at the hospital. After so long spent inside the house, watching the world from afar, it was pleasant to be outside, back in the city that he loved. He could see from the light at the door that the electricity was flickering on and off. He hoped nothing was wrong with the baby, and that they wouldn't have to operate.

Someone rapped on the car window. Abu Laith looked out. There was a Daeshi standing outside, but he looked friendly.

'Assalamu aleikum,' Abu Laith said, as he opened the door on the Daeshi's side.

'Wa aleikum assalam,' the Daeshi replied, smiling. He had a thick village accent and was dressed not unlike Abu Laith in his disguise. 'I'm afraid you can't wait for customers here.'

Abu Laith inwardly rolled his eyes. The Daeshi was such a villager that he thought his Chevy was a taxi. 'My wife is giving birth,' he crowed, adding for good measure: 'She'll be giving me another son for the *dawla*, to make us strong.'

The Daeshi nodded sagely. 'Inshallah,' he said, and started walking back back to the far corner of the hospital, where he had been standing guard in the shadow of the building.

Abu Laith hoped it would be over soon.

Lumia was screaming through her gap-teeth. Her feet pushed against the bed.

'One more,' shouted the doctor, who had pulled back her own *khemmar*. 'Push.'

Lumia gave an almighty heave. She was wrung out as a washing cloth. Then she heard a small cry.

'It's a boy,' said the doctor, rising, the baby in her arms.

Lumia felt dazed, as if she had been in an accident. The doctor had left the room with the baby.

'Where is he?' she breathed.

As she lay there, Lumia took great gulps of delicious, pain-free air, wafting her *khemmar* up and down over her face, ignoring the fact that she wasn't supposed to.

A few minutes later, the doctor came back holding a tiny package of white-swathed cotton. It was crying.

'You're so beautiful,' said Lumia, quietly for once, as she took her son in her arms.

He was no longer than a loaf of bread, only his tiny, screwed up, brown-purple face sticking from the fold of the towels. His cry was loud as a lamb's. His hair was ginger.

'Here you are,' croaked Lumia. 'Welcome, Shuja.'

20

Hakam

Sweaty and tired to the bone, Hakam let the gym doors swing shut behind him and jumped on his bike. He had just finished a marathon workout, and was five minutes away from a protein shake and a shower. Bag sticking to his back, he spun away from the gym and down the road to where the old checkpoint had been. There was a group of people gathering at the side of the road, all men and boys. A man was standing in the midst of the crowd, eyes bound, wearing an orange jumpsuit. In front of him some Daeshis had set up cameras, as if they were holding a press conference. Others were running into the street, grabbing more spectators. Hakam got off his bike and started walking it along the road.

'Come over here!' a fighter yelled at him. 'We're killing an infidel.'

Hakam didn't know what to do. He couldn't bear to stay and ogle this man being executed, but nor could he run.

The crowd thickened, drawn by fascination or disgust. Among them were a few Daeshis and, going by their deeply uncomfortable expressions, some normal Moslawis, forced like him to watch the spectacle. The children gaped at the condemned man. Hakam couldn't take it. Looking around to check no one was watching, he slipped through the audience and walked, then cycled off. As he rode, he felt more and more angry at the public display the Daeshis had made of this man's death. The looks on the children's faces annoyed him. They had been fascinated, unable to look away. After more than a year and a half of Isis occupation, they were used to the brutality; enamoured by it. He didn't want to be like them.

Daesh loved to make a spectacle of violence. Behind his house there was one of many 'media points' that had been set up at major junctions throughout the city. They were, to all intents and purposes, outdoor cinemas built by Daesh, with a projector screen and rows of chairs. On occasion, the Daeshis linked up a laptop and played their propaganda videos on the screen: video from the Paris attacks of November 2015 or GoPro footage from their battles with the Iraqi army.

'You saw them too?' asked his cousin the next day, when he came over, looking extremely worried. Hakam, who had been thinking the scene over all night, had told him about it as soon as he arrived.

'I must have got there after you,' his cousin said. 'They wouldn't let me leave. I had to watch it.'

The man who they killed had been a captured Kurdish soldier, his cousin said. They had made him read out a statement on camera before slaughtering him on the spot.

As time went on, Hakam thought, Daesh was growing more paranoid, and revelled deeper in its brutality. Executions became normal across the city, particularly in the centre, where severed heads were sometimes displayed on spikes. Punishments were doled out in a way that the Isis judges deemed most befitting of the crime. A captured Jordanian pilot was burned alive in a cage in Syria, supposedly because the bombs he dropped had burned others. The video of his death was broadcast on screens around the city. A captured sniper would be killed with a bullet to the head.

But almost the worst was reserved for those Isis militants accused of being gay. Cycling though the city one day, Hakam saw people walking around on the roof of the offices of the National Insurance Company, one of Mosul's tallest buildings and a landmark in the flattish city. A crowd had gathered in the street that ran in front of the building and Hakam knew immediately what was happening. He turned and cycled away, horrified. For weeks, Daesh had been throwing people accused of homosexuality from this building to their deaths. They justified themselves with a twisted re-interpretation of the story of Lot – an Old Testament tale told slightly differently in the Quran.

In the Islamic holy texts, Lot was a Prophet who preached to the men of Sodom and Gomorrah to end their depraved ways – committing acts of banditry and having sex with other men – and embrace God. But they rejected him, and as punishment God turned their cities upside down and rained down stones upon them. It was the act of turning upside down that inspired the Daeshi's punishment – flipping men through the air just as God had inverted the two cities.

In this climate of brutality people closed themselves off. Talk between friends drifted always on the surface, never commenting on the inescapable fact that their city was run by a violent cult. People did strange things they thought would make them safe. One of Hakam's friends volunteered to join Daesh. He was a chemistry student, and had been afraid that the militants would find out that he had once worked as an election observer for the Iraqi government. A few months later, they found out, and killed him anyway – two bullets, one in the chest, one in the head. He had a one-month-old daughter.

The Daeshis knew so much. Though a lot of them were ignorant, village-idiot types, their leaders certainly weren't. When they took the city, they had stolen or hacked their way into government and army records listing everyone who had worked for them. They had intelligence records going back decades, and added their own lists to them. Europeans, Arabs, Asians: their intelligence department was made up of smart, brutal people from across the world. They could listen in to phones, Hakam feared, or restore deleted messages.

When he spoke to his brother Hassan in Pennsylvania, he made sure only to talk about safe subjects, and tell him they were all fine. The internet was so sporadic that it was hard to communicate at all. When Isis had come the government had closed down the mobile network coverage and, later, the city's broadband network. Instead, Moslawis resorted to unreliable satellite internet connections.

For the first time in years, Hakam had to spend hours struggling to connect to the internet. It was one of the many ways, it seemed, that they were travelling back in time.

21

Marwan

THERE WAS ONLY REALLY ONE REASON WHY MARWAN enjoyed working in the zoo. He liked animals, and he desperately needed money, but he was above all trying to meet girls.

For a young man of limited means in Mosul, that was almost impossible to do. Marwan's parents were poor, and they didn't like him very much. Getting married was expensive: first a dowry had to be paid, in the form of gold, jewellery or cash, then the wedding itself – hundreds of friends, neighbours and hungry acquaintances eating and dancing for three days. Then, finally, you had to buy your own house and support your family. Marwan had no money, no family inclined to help him, and was as a result starved of female company.

There was no chance, he would think gloomily, as he watched the families walk by the zoo enclosure, that he would ever be able to find a wife. But he did hold out hope that something more short term might crop up. He knew there must be girls up for some fun, but this was a difficult environment for casual sex. Daesh would certainly kill him if they thought he had lusted after one of their own women.

Even a regular Moslawi woman might be killed by her family if she was caught having an extra-marital liaison with Marwan. 'Honour', as they thought of it, was a perishable commodity held by women, and its loss a greater disaster than almost anything else that might befall them.

But Marwan was bored, and he had watched a lot of romantic films, and he could dream. Maybe there were girls who felt the same, he thought. So far, his attempts to talk to women at the animal enclosure hadn't gone very well. They were too scared of him, or their family, or Daesh, to respond. He wished he was better looking. Marwan hated his dark skin, and, when it really came to it, couldn't really countenance the idea that a woman might ever love him.

It was in the midst of such reflections that two women came to the zoo enclosure one day. Marwan jumped up and went to the entrance to sell them tickets. He had become pretty adept by now at telling what was under those black cloaks, and he saw immediately that one of the women was young, the other older.

'Assalamu aleikum,' said the older woman, and he greeted them in turn.

'How much is it to get in?' asked the younger one, and Marwan almost fell over at the musical beauty of her voice. He couldn't speak for a moment. He just stared, hoping she would speak again.

'Five hundred dinars,' he said, finally. 'Each.'

The young woman giggled. 'We don't have any money,' she said.

'Oh,' said Marwan. 'I mean, you can go in for free.'

The women laughed. 'We were only testing you,' said the young one, handing him 1,000 dinars. 'We've got money.'

Marwan laughed, still feeling dizzy at the sound of her voice. He opened the gate and showed them in. He would give them a guided tour, he thought, rallying himself. That would impress her. 'This is the monkey cage,' he said, with a proprietorial air. 'There are several monkeys here. One of them is Nusa, you see, the red one.' He opened the cage and walked in – gasps from the women – and let Nusa take some sunflower seeds from his hand as the other monkeys scampered around him.

The young woman laughed. Marwan swelled. He thought no end of himself – a professional monkey trainer and seducer. In a stroke of inspiration, he snapped his fingers and clicked his tongue for the baboon, who he had been trying to train to come to him when called. The baboon heard him and perked up, swinging towards him through the cabin and landing on a branch nearby.

'How do you know how to do that?' the girl asked in her beautiful, and now admiring, voice.

Marwan tried to look casual. 'I'm an animal trainer,' he said. 'That's what we do.' Marwan turned and smiled at the girl, full blaze. Then he saw her put her hands over her face and scream with laughter.

'What is that monkey doing?' she shouted. 'Look, look.'

Marwan spun to look at the baboon. He was dancing around on a branch, holding on by one hand, the other hand locked around his genitals, which he was swinging around in a triumphant manner.

The women screamed and ran away, hands over their faces. Marwan, caught between amusement and anger that his potential girlfriend had been scared off, gave the baboon a kick, and it ran over to the other side of the cage, hooting. 'Stop ruining things,' Marwan shouted. Flustered, he slouched off to the corner of the enclosure. Another one gone. He would never find a girl at this rate, he reflected, furious at the monkey.

'Excuse me,' came the lovely voice again. 'Could you fill up my water bottle?' The girl was standing back at the entrance to the enclosure, holding out a water bottle with her black-gloved hand. She was by herself.

Marwan ran over, heavy with relief. 'Sure,' he said, beaming. He took the bottle, which had an inch of ice at the bottom, and filled it up at the tap in the enclosure.

'Thanks,' she said when he gave it back. 'What's your name?'

'Marwan,' he said. 'What's yours?'

'Heba,' she said. Marwan thought she sounded pretty. 'You're really confident, aren't you? Just like that monkey.'

Marwan was surprised. This was not how nice Moslawi girls talked. He didn't know what to say, and she was just standing there, looking at him, with her water bottle in her hands.

'I think love is sent from God,' he finally choked out. 'Because I am in love with your voice.'

It was the right thing to say. In a place where courtship was usually short and condoned by parents, declarations of love were made quickly and passionately. The girl laughed again, looking to check no one was watching. 'I want to see you again,' she said. 'Can I come back here?'

'Whenever you want,' said Marwan. 'I'm always here. Please come, all the time.'

'OK then,' she said, turning around to go. 'See you.'

It took a while, Marwan later remembered, before he managed to calm himself down. There was a girl. She liked him, and was coming back to see him. She might even be pretty. He'd said he loved her voice, which he supposed was the right thing to do. Unlike his romantically minded friends, he was certainly not the sort of person who fell in love at first sight, but that was just the way people talked here when they were seeing each other. He hoped she'd come back.

Almost exactly a week later, Marwan was standing by the entrance to the animal enclosure when he heard Heba's voice again.

'Assalamu aleikum,' she said, and giggled. Marwan, who had been daydreaming about what her face might look like, was extremely pleased to see her.

'Wa aleikum assalam,' he said, trying to look casual. 'How are you?'

'Are you going to let me in?' she asked. He opened the gate. It was still early in the afternoon, and there was no one else around. 'I'm here with my parents,' Marwan remembered her saying to him in that musical voice. 'We're going to have to be quick. They think I've gone on one of the rides.'

The park was beginning to fill up with women in identical black coverings. As far as her parents knew, Heba could have been any one of the girls on the carousel. It was, Marwan thought, the perfect cover for their affair.

'I love you,' he said, after they had chatted for a few minutes. She lived on the outskirts of Mosul and was just finishing school. She was eighteen years old, a bit younger than him.

'I love you, too,' she said.

Neither of them really meant it, he reflected after she had gone, leaving him feeling strangely light and satiated. He was mainly just hoping to talk her into bed. But she seemed fun, and not like any of the other girls he had – albeit briefly

– spoken to. She was easy to talk to, and she was kind. She promised to come back next week.

That Friday she came a bit later, just before the call to prayer, when the sun was about to set. It felt different, this time, more urgent. Marwan had been shaking to see her. He wanted, more than anything, to see her face. Pale, he thought, and beautiful, probably, but beneath his romantic and lustful inclinations there was a niggling fear that she might only sound beautiful. He wasn't sure what he would do if her face was ugly, and it was driving him slowly mad.

'This time I need to see your face,' he said hotly, as soon as she arrived. 'I need to.'

She had brought him a sandwich, as she had done before, and handed it over with a black-gloved hand. 'No,' she said. 'I'm too scared. There are Daeshis everywhere here.'

But Marwan was determined. 'I love you,' he said, not really meaning it. 'I want to marry you when this is all over. Please, let me just see your face for one second.'

'There's nowhere to show you,' she said. 'People will see.'

Marwan looked around the enclosure. Over by the corner there was a tree which spread its branches far out along the edges of the animal enclosure. The back of the tree faced the wall, and the view into the park was blocked by branches on two of the other sides.

'Come here,' he said, and he ran over to the tree, Heba tailing behind him.

'Someone will see.'

'Come here. Please.'

Heba waited a moment. Then she ran forward under the tree, just inches from Marwan. He wanted to touch her, but there was no way he could do that, and he let his hands hang by his side.

'Ready?' she asked.

'Ready.'

And Heba flipped her face veil up.

'You're beautiful,' Marwan breathed, and he really meant it. She had pale and perfect skin, big eyes rimmed with kohl and bronze, and she was smiling at him like she was honestly happy to see him.

Then the face veil was down again and she was running through the zoo, and through the gate over past the bear's cage. 'See you next week,' she called.

Marwan was still standing by the tree, his heart full. He was in love. Really in love, not like he had been before, when he was just trying to charm her into bed. He thought about her face, and how perfect it had been, and tried as hard as he could to sear it into his mind so that he would remember it for ever.

A pattern quickly emerged. Each Thursday or Friday Heba, who had convinced her family that a trip to the park was the ultimate weekend activity, would come for a picnic with her parents and siblings. They would spread out on the grass with their sandwiches, sitting on mats. After a while, Heba would take one of the sandwiches and tell her family she was going to go on the rides. Then, blending into the hundreds of other black-clad women, she would sneak over to the zoo enclosure and see Marwan. He would eat his sandwich and they would talk – about when they would get married, about their families and what they wanted their lives to be.

Marwan, who knew that his parents would never pay for a dowry, had asked Abu Laith if he would help him. Abu Laith was very pleased at the prospect. He thought Heba sounded exactly like the sort of girl who would whip Marwan into shape.

'I don't understand how you can love me,' he told her one afternoon. 'I'm poor, and I have such ugly black skin.'

'Of course I love you,' she said. 'You're beautiful.'

One weekend, she had bought him a gold necklace with the letter H hanging on a pendant. He had worn it under his shirt, close to his chest. Another time, she had bought him a blue shirt and green trousers, full-length and tight, a style Daesh had banned. Marwan didn't come from a family where anyone had ever really cared about him, or given him presents, and the idea that someone beautiful and kind could really love him was sort of staggering, and it confused him.

Gradually, after a few months of seeing each other they began to broaden their horizons. Now, they no longer just met in the animal enclosure. Walking together like husband and wife, they went on the Ferris wheel, sitting next to each other and looking out over Mosul, grey and quiet in the early months of 2016. They went to feed the birds in the pond. Because Marwan worked there, they went on all the rides for free, which made him feel important, and impressed Heba – he thought – no end. Her parents might have been next to them, but they would never know it was her there with the young man with the dark skin.

'When are we going to get married?' he asked one day as they sat by the pond.

'After this is over,' she said, as usual.

'But why not now?' he asked. 'I love you.'

'I love you too,' she said. 'But we have time.'

'Unless I find another girlfriend,' he said, teasing her, and she shoved him in the ribs so hard he was winded.

22

Imad

2003

It started with a stomach ache. A sharp pain that made Muna fold in half sometimes. She had, she told Imad, been feeling tired for a few weeks, but had thought nothing of it. She was almost always very tired. Oula, her youngest child, was still in her crib. Mohammed, the second youngest, was just learning to walk, and was constantly getting into trouble – falling onto the pigeons with their sharp beaks and flapping wings, or eating things he found lying around.

Another six children, the oldest almost twenty, were spread around the house and Mosul. Another, Qusay, lay buried – he hadn't lived long enough to crawl. Like many Moslawi women, Muna had been raised to think that having many children gave security and wealth. But even before the stomach ache, her body had been breaking, sagging together like a tent after a rainstorm.

Then the stomach aches started. She complained, and Imad told her to go to the doctor, but she refused. It was a while before she finally went to the hospital. The doctor said it was 'women's problems', gave her some pills and told her to go home. Her illness didn't get any better.

A few months after the stomach aches began, Imad took his wife to the hospital. He walked into the doctor's office alone after Muna had been examined. In Iraq, bad diagnoses were often given to family members rather than the patient, because of fear that the shock would kill them.

'It's cancer,' the doctor told Imad. 'Pancreatic cancer. But don't worry' – he looked at the dumbstruck man – 'there's a decent chance she'll make it.'

When they got home, Imad promised Muna that they would get her better, whatever it cost.

It didn't take long before she couldn't clean the house, or feed herself, then – finally – move at all. She shrank, wasting away until her body was little more than skin stretched over jutting bones. The children were scared of her, no matter how much she tried to care for them.

They were scared a lot. The Americans had invaded Iraq in 2003, the year Muna got sick, and now they drove through the streets of Mosul in their armoured vehicles, guns at the ready. They were battling a Sunni insurgency, dominated by al-Qaeda, that was tearing the city apart with suicide bombs and attacks on US troops and government forces.

Many in the city hated the Americans. Most Moslawis were Sunni, and had good lives under Saddam, who treated them well at the expense of the Shia majority. While many in Mosul suffered under the sanctions in the 1990s, the threat of starvation or death was not as great as it had been in the Shia strongholds in the south.

No one in the family cared much about politics. Imad, who

had seen his friends and his brother killed in the Iran–Iraq war, was – on balance – happy that the Americans were there. He said as much one day when they came to his home. They were combing through every house on the street for weapons after suspected al-Qaeda militants had attacked US soldiers in the neighbourhood.

They knocked on the door, and Imad let them in. 'Welcome,' he said, as they filed past him. Their leader was black, like the Sudanese people and the black Iraqis he had seen in Baghdad.

Their translator addressed him. 'They're asking if one of your ancestors was European,' he said, indicating Imad's shock of ginger hair.

'No, I'm Moslawi,' Imad replied, proudly. 'It's only my tribe who have this hair.'

Weapons at the ready, the Americans looked through the house, and saw Muna lying in a cot in the living room. They asked if there was anything they could do, and Imad said he didn't know – that she was sick with cancer. Their leader called their base and asked for medicine, which Imad thought was decent of him. But there was nothing they could really do, and they were soon gone, leaving the family behind.

At 08.39 a.m. on 23 January 2007, Muna Fadel Said al-Shamaa died, leaving a husband and eight children behind. All of the men in the neighbourhood came to her funeral that afternoon in the Jadida district of Mosul. The children all stayed behind in the house with the women. A car bomb went off in a nearby neighbourhood as they buried her.

Imad was worn as an old dishcloth. He had stood at the mosque as the imam performed his prayers, and felt as if he might blow away. He had been so angry – more furious each time the doctors hadn't cured her. All the money he had, he poured into the wards and the clinics and the machines that, somehow, seemed to make it all worse.

When she had died, Imad had told the children straight. There was no sparing her death, which had marched towards them clear as fate. They had cried, and now she was gone, and they didn't have a mother any more.

As the imam spoke, Imad could hear his neighbours whispering. He knew what they were saying. They were pleased to have been right. For years, they had complained about their infidel neighbour who drank whisky and kept filthy dogs that barked late into the night. Now God's anger had struck his wife and killed her. It was no one's fault, they had told Imad many times, but his own.

When his wife was alive, Imad had always laughed off their comments. They were suspicious peasants, he said, who would believe anything anyone told them. His wife had cancer because a tumour had grown in her pancreas and spread through her body.

Now, as his wife was lowered into the ground in a white shroud, he started to think they were probably right. He began to cry then, great shoulder-heaving sobs that shook him from the core. He was desperate, trapped and looking for a solution. He found it, in a moment, inside himself. Her death was his fault, he knew, and he would spend the rest of his life repenting. He would never bring the wrath of God down on his family again.

From that moment, Imad became a holy man. Every day, he would pray five times, raising his hands to heaven, slowly getting to his knees, touching his forehead to the floor. He prayed to God to save his family, and to take his wife into heaven. He used a large chunk of his earnings from the mechanic shop to build a mosque just yards from his house by the park.

When he saw someone drinking – even his old friends, with whom he had raged through Mosul for decades – he would tell them it was forbidden. The neighbours, who had whispered about him for so many years, were thrilled. Imad's beard grew long, though he never trimmed his impressive moustache, as the Prophet Muhammad had. After evening prayers, he would sleep in the mosque he had built, to guard it from thieves.

If the children woke up in the night, they would gang up and run over to the mosque, clambering over the fence and knocking on the window to wake their father up. He would let them in, and they would sleep in a pile on the floor until dawn prayers, when he would make food for the worshippers, who had radically increased in number since word got round that the new mosque down in Hayy al-Nur gave out free breakfasts.

In the evenings, on occasion, his old friend Sheikh Hassan Ali Beg would come to plead with him to see sense. 'The mullahs are hypocrites,' he would say. 'You know they are. Why are you abandoning your friends?'

Imad didn't believe it. He was still feeling empty after Muna's death, and the belief that he might be saving his children by becoming a holy man made him feel a little more positive about the future.

He was constantly disgruntled, though, by how little faith his children seemed to have in his transformation. Lubna and Laith seemed to find his rants about the nature of God quite funny, and the small ones openly laughed at him. Sitting with him in the mosque, they told him again and again that it wasn't his fault, and that the mullahs were liars.

The first sign that Imad's transition to orthodoxy might not last very long came one sunny morning when Abu Saad, one of his neighbours and a man of impeachable piety, asked if he would fix the speakers on the mosque to make the call to prayer even louder.

'Of course,' Imad said, and climbed up on the mosque roof. He leaned a ladder against the turquoise metal struts of the minaret, and began to climb. He was working at the wires when the ladder collapsed under him. He fell, straight down, over two stories into the courtyard of the mosque.

'This is ridiculous,' he thought, as he looked down at his foot. His calf bone was sticking straight out through his skin. 'What the hell am I doing?'

It didn't take long before, on a trip to Baghdad, he lit out for a rowdy place of ill repute, where someone shoved a cold beer into his hand.

A wave of happiness flowing through him, he took a great swig. 'I'm back,' he shouted. 'Anyone got any whisky?'

23

Abu Laith

THE RUMOURS THAT HAD BEEN SURFACING FOR SOME TIME came to a head one autumn morning in 2016 when Abu Laith was – not for the first time – standing on the sofa, scrabbling around in the hollow cornicing on the wall in the corner of the living room.

Under a bunch of fake flowers, some gravel and a couple of stray buttons lay a silver Nokia of uncertain vintage, fixed with black duct tape and wrapped in a plastic bag. It was unusually quiet in the house. The children had been shut into the garden, and ample consequences threatened should they look into the living room.

Abu Laith pulled the plastic bag out of the cornice. He took out the phone and placed the fake flowers back in the hole.

More than a year before, Isis had pasted shop walls with another edict: anyone caught with a SIM card would be punished. Like many others, Abu Laith had taken the precaution of hiding his phone, restricting it for emergency use. Now was such a time.

For a few weeks, they had been expecting to hear that a campaign to liberate the city was about to begin. It would be the American-led coalition and the Iraqi army, the family had heard, who would advance on Mosul and flush out Daesh. During the rare times when the electricity was on and they could get satellite reception, they tried to watch the news on TV – sound turned right down and curtains drawn, so that no one could hear them. But they just didn't know whether it was true. The government had lied to them before. Daesh said that the army would slaughter Moslawi civilians if they came. Abu Laith and Lumia, as they watched the news, children locked outside, would not allow themselves to believe that the end could really be coming.

If the army was going to try to take Mosul, Abu Laith had reflected, it would mean war. Real war, like he had seen on the Iranian border, buildings turned to dust, children blown apart by shrapnel. He had just spent two years confined to his house, his family splintered and living in constant fear. That was bad enough. Death, the possibility of a bomb falling on the house, or his children being blown apart, was quite another thing. The current government had never cared about Mosul, Abu Laith later remembered thinking, so there was no reason why they would try hard to avoid civilian casualties if they did come. He wanted Daesh gone, but he did not want his family to die in the process.

The Daeshis seemed to think something was happening. The gunmen in the zoo had grown so paranoid that they had started arresting visitors on the flimsiest pretexts. Marwan still handled the feeding of the animals. He was so surly and preoccupied that anyone would think twice about bothering him.

Abu Laith had ruminated on the possibilities. If the army came, they'd see the Daeshis in the zoo, and bomb it without a second thought. Zombie and Lula would die. He needed to tell the army that the animals were there. He needed to understand what was going on. He needed, in short, to make a phone call.

'I'm going to the roof,' Abu Laith told Lumia, over baby Shuja's wails. 'I need to make a phone call.'

'You can't,' she said. 'It's dangerous up there. And you'll get caught.'

Mobile signals in the city had been blocked for years, since the Iraqi government cell companies had been ordered to stop operating in Daesh areas. But everyone knew that you could get a signal if you were high enough, or on the edge of Daesh territory.

'I'll be back in a minute,' he said. 'Don't worry.'

Abu Laith snuck up the dusty stairs. He pulled open the attic door and went onto the roof.

Crouching low, he took out his phone and dialled a number.

A husky voice answered.

'Baba?'

It was 6.30 a.m. and Abu Laith's daughter was half-asleep.

'Dalal,' exclaimed Abu Laith. He admired his oldest daughter with an uncomprehending pride, this businesslike woman who worked in military intelligence and who had rescued three of her siblings and taken them to Baghdad with her when Isis first arrived. Lubna was now working in a beauty salon and Mohammed and Oula going to school. For them, it was an adventure. For Dalal, it was duty. She had not flinched at being asked to house and care for her siblings.

Crammed into the tiny apartment in Baghdad, they talked endlessly about their family back home. Dalal knew very well that Moslawis who had relatives in the army were being targeted by Daesh. But she had never worked in Mosul, and hoped they wouldn't make the connection.

As often as he could, Abu Laith would call and tell Dalal the latest news about the zoo. Every time, he risked his life and that of the family to speak to her. He always told her he was going crazy because he couldn't see Zombie.

This time, he sounded more hopeful. 'Is it really true?' he asked. 'Is the army really coming?'

'I'm outside Erbil with the army,' she said. 'The operation is just starting. We'll be there in a couple of days, God willing.'

A couple of days. Through the euphoria that pumped through him, Abu Laith couldn't quite believe it. He was still uneasy: the army didn't know that there were animals in the zoo. They would target the first Daeshi that they saw there. He knew, with the certainty of an old military man, that the Daeshis would use that patch of open ground. He had to protect the animals.

'I'm going to tell you some things,' he said. 'I'll tell you information, and you tell the army.'

'I am the army,' she said, faintly exasperated. 'You can tell me.'

'You have to promise to call your commander.'

'Fine, baba.'

Ten minutes later, they rang off.

She later told Abu Laith that she had called her commander, and passed on the message from her father: 'There are three lions, two bears and many other animals in Mosul Zoo. The people on the road next to it are civilians. The fighters are cowards who'll run as soon as you come. Please don't kill us.'

Abu Laith had been pleased with its brevity.

•

Long after the line had gone dead, he was still standing on the roof. Behind him, towards Erbil, the suburbs of Mosul stretched out for miles towards Gogjali – grey and dusty in the morning sunlight. All along the way to the outskirts of the city there were Daesh checkpoints, and Daesh families living among ordinary Moslawis. The jihadis had known for years that the army would eventually attack, and prepared their defences. It did not seem likely that the army would be able to enter the city in a mere matter of days, as Dalal had suggested.

He calculated the amount of food the family would need to survive a siege: working it out in sacks of flour, beans and rice. They had water from the well in the zoo, and could bake bread on a stove on the roof, if they needed to.

War was coming, and possibly, freedom. But when he thought about the children, and Lumia, and the new baby squalling day and night, the fear stuck hard through the excitement.

For now, he would not tell Lumia or the children, he decided. His offspring were the biggest gossips in the neighbourhood, and they would – he was entirely certain – tell everyone from the Daeshis to the neighbour's cat if they knew the army was coming and, worse, that they had a sister in military intelligence.

He would tell Lumia later. Sometimes he worried about her nerves. It could be a false alarm. Dalal might have misunderstood her commanders.

But when he came into the kitchen, he couldn't help himself. Looking around at his wife and children, who might soon be free, he started laughing a deep, joyful belly laugh.

'What is it?' asked Lumia.

'Nothing,' he said, grinning. 'Nothing at all.'

24

Imad

THE LIVING ROOM WAS THICK, AS IT SOMETIMES WAS THIS time at night, with the smell of whisky and the distant snores of children. Next door in the hall by the kids' room, the TV blared – tuned almost permanently to National Geographic. Imad, now that his children were in bed, was lying prostrate on the floor, sipping from a glass of Black Jack, the world's worst whisky.

Over the speakers, Umm Kulthoum's voice drifted through the violins accompanying her. Imad was very specific about the choice of music. He listened to the Lebanese diva Fairuz in the morning. But the evening was for the great Egyptian singer Umm Kulthoum. As he lay there, half-cut as he was about once a week, he thought about Sara, the woman he should have married, but never did.

There was a knock on the outside gate, and Uday, Imad's brother, shouted a greeting. Uday didn't drink, but was wary enough of his older brother's temper not to complain when Imad did.

He could come in, Imad decided. 'Welcome,' he cried.

His brother opened the living-room door and looked down

at him. Uday's hair was as red as Imad's, and he had the same rough-hewn nose, which wrinkled when he saw what Imad was doing. 'Are you drunk?' he asked.

'Barely,' said Imad cheerfully, and sat up. 'How can I help?'

'If you're drunk, it can wait,' Uday said. 'This is really important.'

Imad jumped up. 'Come on,' he said. 'I swear I'm not drunk.'

Uday eyed him critically, then walked into the living room. The brothers sat down. Imad was surprised to see that his brother was looking genuinely worried.

'If I tell you something,' Uday asked, 'do you promise that you won't get angry?'

Imad considered him. He did have a short temper, but only really lost it when when someone tried to con him out of money, disrespect him or lecture him about religion. He did want to hear what his brother had to say. 'I promise,' he said.

Uday stood up and walked over to the door. 'You can come in,' he said.

Imad looked up. Standing in the doorway was Maan, Sara's brother. It had been twenty years since he had last seen him. Maan and Imad had been best friends and drinking buddies until Sara had rejected his proposal. Then, Imad had declared a blood feud against Sara's family – vowing to kill any of her male relatives who came to Mosul.

With the whisky flowing through his blood, Imad felt the old fury come back – years of abandonment and self-pity after Sara's family had rejected him. 'Why did you come here?' Imad asked. 'Why did you break the peace?'

Maan looked defeated. 'I missed you, brother,' he said. 'I didn't want there to be bad blood between us.'

'Of course not,' Imad said. 'But what else could I have done? You insulted me so deeply. I thought we were like brothers, but you deceived me.'

Maan looked confused. 'How did I ever deceive you, brother?' he asked. 'If a deception was made, it wasn't by me.'

'You know what you did,' Imad shouted. 'Why did you turn down my proposal?'

'You never proposed to Sara,' Maan said. 'She was devastated. And then you sent a message that we could never set foot in Mosul again.'

Imad stared at him. 'But I sent my parents down to ask,' he said. 'They came back, and they said she had turned me down.'

'They were lying,' Maan said. 'Sara always wanted you. She still cries for you.'

Imad slumped back onto his blankets, thoughts whirling. Then a glint lit in his eyes. He jumped to his feet, nimble despite the Black Jack. 'Take me to her,' he bellowed, wild as a bull. 'I need to be sure.'

It wasn't more than ten minutes before Maan's feeble protests – it was late, they were drunk, Sara was married to someone else, they hadn't seen each other for twenty years – were bulldozed by Imad, and the three men were in Imad's black and silver Chevy, roaring through Mosul.

'So her husband's in jail,' cried Imad, thrilled to the core. 'What a stroke of luck.'

Sara's husband was a few years older than her, and did not like animals. He had been jailed for stealing 25 million dinars from the Pepsi factory where he worked as an accountant. He hadn't actually stolen the money, his family thought, but that didn't make any difference to the judge, who sentenced him to twenty-five years in jail.

Sara had been left at home with her three children, still miserable, and it had seemed she would remain there until her husband was released. Imad, however, had other plans. As the car rolled through the plains south of Mosul, he fantasized about seeing his nubile young Baghdad temptress again. It had been a few years, of course, but his feelings hadn't changed, and he did not for a second countenance that hers might have, either.

They arrived into Baghdad at 3.30 a.m., pulling into the dark street where Sara and Maan's extended family lived.

'Stay here,' said Maan, as he got out of the car. He was, Imad noticed, beginning to have some doubts about this mission. After Imad had declared war on the family, Sara had flown into a rage, followed by deep depression. Bereft of any other evidence, she thought that Imad must have found someone else, a suspicion confirmed when she heard of his wedding to Muna.

In an moment of unusual pettiness, Imad had arranged that an announcement of his wedding be sent out to everyone in the extended family, on pieces of ornately decorated paper, accompanied by some sweets. Sara had broken down when she saw it, but she still believed that he would come for her, until she finally gave up.

Imad waited in the car, feeling very pleased with himself. Sara wanted him back, he knew. He heard the door to the house open, and someone walking out.

Then she was there, her long hair down and her eyes staring, frozen to the spot in her blue and white pyjamas. Her face lit up, and she ran to the driver's door, wrenching it open.

A flurry of hair and arms and then Imad was hugging her. They were both crying, great sobs that shook them.

'I burned for you every day,' he said. 'I couldn't eat if I wasn't remembering you.'

'I'm here now,' she said.

25

Abu Laith

IT DID NOT TAKE LONG FOR THE HOPE OF IMMINENT liberation to wear off. A few days after Abu Laith had spoken to Dalal on the phone, the family had seen on the TV that the start of the operation to liberate Mosul had officially begun. There was rolling footage of Isis car bombs and improvised explosive devices (IEDs) being blown up by the army on the plain near Gogjali. But the army was moving slowly, and the noise of the battle only reached the family from a considerable distance. There was now no phone signal at all, and it was too risky to spend much time on the roof.

In the few weeks since October had turned to November 2016, the shelves in the shops had emptied, and Isis grown more deadly than ever before. The advancing forces had blockaded the city so tightly that not even a rat could squeeze out of Mosul. Bags of flour, when they could be found, cost 4,000 dinars each – up from 250 a bag that summer.

One afternoon, when the cold wind had begun to bite and the ostrich in the zoo was shivering, Abu Laith stood by his empty fridge, deep in thought. He had bolted down his lunch because he had suddenly decided to do an inventory of what was left in the house for the animals, and Lumia had shouted at him, again. Around him his offspring ran through the house, screaming and shoving. Pigeons flapped around them. A dead sparrow hung mummified in a spider's web on one of the outside walls.

Their lunch had been uninspiring. Today, as usual, it had been rice, boiled then doused in oil to make it more filling, with some pickles on the side. None of them had seen a tomato for weeks. Lumia's hair was growing grey, her round cheeks starting to cave in. Baby Shuja was lying in a cot in the bedroom. He cried even more than her other children had. Though this meant she couldn't sleep at night, Lumia knew it was probably a good thing. There was not much, she thought, that would convince her to go back to that hospital.

Sometimes, Lumia wondered what she would do on the day that her children were truly hungry, and she had no food to give them.

'The monkeys would have liked some of those pickles,' said Abu Laith despondently.

Lumia snapped. 'Why do you care?' she said. 'Just eat the food and be happy to have it.'

'You're human, and you have a tongue, so you can say that you're hungry,' Abu Laith said. 'But the animals can't. And they get hungry like humans do too.'

Lumia, who had been angry about this for a while, reared up at him. 'Our neighbours are starving,' she said. 'And whenever you have chicken or honey you never give it to us, or to them, but only to your stupid animals.'

'They're helpless,' said Abu Laith. 'They'll die if we don't give it to them.'

All over Mosul, those who could were starting to stockpile

food ahead of the coming war. Though they might not like the militants, locals still feared the coming battle. Lumia was no different. She wanted Isis gone, but she had lived through the American-led invasion already, and knew what war looked like. There would be hunger, violence, shelling. She knew that civilians would bear the brunt. Even worse, they had heard on the TV that American planes would be bombing them.

Abu Laith and the children were no less worried about food than Lumia was. But their main concern was for the zoo animals. They were never so hungry that they couldn't be distracted by one of their moaning siblings, or a quest to discover new bugs. The animals, on the other hand, were bored, starving and apathetic.

For months, there had not been quite enough food for Lula, Warda, Zombie, Mother or Father, nor for the monkeys and dogs. The peelings and eggshells that the children brought them as snacks from the kitchen were leaving them thinner than ever.

Abu Laith sat down and did some maths. Each lion, he calculated, would ideally have 10 kilos of fresh meat a day. Lula the bear put away at least 5 kilos of bananas and 10 of river fish. The monkeys needed bananas, when they could get them, as well as hard-boiled eggs. A guinea pig could never be happy without lettuce.

He thought of the dogs. They were so hungry they had stopped barking, he had heard. A shame on the zookeeper who fed them.

With the prices rising, his savings wouldn't raise even a quarter of the food needed. He needed more money, quickly. But his phone barely worked, and he wouldn't risk using it to try to contact the zoo's owner, who lived in Erbil and didn't particularly seem to care what happened to the animals.

He sent for Marwan, who confirmed that Zombie, Mother and Father were in a bad way.

Something drastic had to happen if his charges were going to live. Abu Laith thought for a while, then rounded up the kids with a yell. Clattering and shouting, they ran into the courtyard and sat down.

'I think the animals are going to die,' Abu Laith announced. 'Unless we work together.'

The children quietened down, like birds settling into a nest.

'You'll have to go hungry sometimes to feed the animals,' Abu Laith said. 'If there's one day when we can buy honey, we'll give it to the bears instead of eating it. If there's meat, we'll feed it to the lions and eat rice for ourselves instead.'

Abdulrahman piped up. 'I'll give my meat to the animals,' he said.

Luay spoke. 'I'll ask my friend at the shop to give us his scraps.'

Minutes later, they had a plan. Abdulrahman was busy at the other side of the courtyard, tying a rope to a basket so that he could drag it behind him. Luay had run to the shop to kick some sense into his friend.

By lunchtime that day, an unruly convoy set off from the house by the zoo. In tow were Abu Laith's children and their associates, as well as a dog. Abu Laith, cursing his bad luck, stayed at home, out of reach of Abu Hareth.

When the group reached the main road, they scattered along their chosen routes. Some ran off to door-knock at the richer houses, asking for scraps. Others headed for the restaurants closer to the centre of town.

Abdulrahman, dragging the basket behind him, aimed for a comprehensive tour of the nearest vegetable market – half a dozen carts that stood at a dusty junction a few hundred feet from the zoo. He promised himself that he wouldn't come back until his basket was full.

He walked round the corner and stopped at the first stall. 'Excuse me,' he said. 'I work at the zoo. We have three lions and two bears and they're hungry. Do you have anything I can put in my basket?'

'There's a zoo here?' said the man by the cart, not unkindly.

'Yes, it's over there,' said Abdulrahman, pointing behind him. 'There are three lions and they're all related. There are two bears too. My dad says they need to eat every day.'

The man looked at him. 'And I imagine that you're here because your dad can't afford enough food for three lions and two bears?'

Abdulrahman nodded mutely. The man dipped his head down behind the cart. After a minute, he stood up and started throwing handfuls of vegetables – some old and squashed, others with crisp new leaves – into Abdulrahman's basket.

'Thanks,' Abdulrahman said.

Flushed with triumph, his basket full, he grabbed his rope and spun into the dusty afternoon.

The others had been just as busy. By the time they reconvened at the zoo later that day, they had gathered a remarkable spread.

In the courtyard by the lion's cages there was a pile of bread, most of it stale as rusk and liberated from the bins by the bakery. Next to it was Abdulrahman's basket of vegetables. Geggo, who had been knocking on the doors down the road, had been given some old rice. Luay had scavenged some scraps from local restaurants. Best of all, one of the searchers had found some bones over by the old sheep market that still had some meat left on them.

Abdulrahman held out a banana through the chicken mesh that covered the monkey cage. Though he loved the monkeys, he didn't trust them. He had seen them steal food from the other animals more times than was reasonable. Hopping forward, one of them took the banana. As an experiment, he pushed a selection of vegetables through the bars: carrots, cabbage leaves and onion.

The monkeys hopped over, curious at the new offerings. One of them picked up the carrot and took a bite. He seemed to like it, because he started scarfing it down, bits of carrot all over his face. The others pounced on the cabbage. When the blur of skinny brown limbs cleared, the onion was still lying on the floor of the cage. Abdulrahman made a mental note to tell his dad that they didn't like it.

Dragging his basket behind him, he made his way around the corner to the next cage, hoping to interest the peacocks in some half-rotten apples. But the cage was empty, and the chicken mesh on the front had been ripped down. Abdulrahman's stomach dropped.

'Help,' he shouted, running back to the others. 'Someone has stolen the peacocks.'

Abu Laith was sitting on the sofa in the living room, working out a schedule for the children. They would have to feed the animals in the evening, he decided. They would spend the afternoons finding food, and then bring it to the zoo afterwards.

Lumia had been angry when he told her that the children would be giving up their food for the animals. But he was very proud. There was nothing that made him happier than knowing that his children were good people, who did the right thing when there were hungry animals to be fed.

He heard the courtyard door swing open, and Abdulrahman ran in. 'Baba,' he shouted. 'Daesh have stolen the peacocks. The cage is open. They're gone. I only just saw them.'

'What?' roared Abu Laith. 'The thieves. I'm going to kill them.'

Lumia rushed in to the living room. 'Oh really?' she said, voice acid. 'You're going to take on Daesh by yourself because they've stolen your stupid birds? You have fourteen children, and Daesh has put a price on your head, and you're going to risk their lives and yours to get revenge over some birds?'

Abu Laith weighed his options, and backed down. 'Sorry,' he muttered. 'I won't go after them.'

'Too right you won't,' said Lumia.

26

Marwan

MARWAN STOOD BY THE TAP, RED HOSE IN HAND, LETTING the water pool over the cement floor by the ostrich cage. It was around 3.30 p.m., and the afternoon was still hot. He was attempting to clean the cages, though he couldn't get inside them to do so, and just spraying water on the floor seemed to make little difference. There were no paying customers at the zoo any more.

Every day planes had been criss-crossing unseen, high over the streets of east Mosul, dropping bombs that sent up plumes of black smoke drifting on the horizon for hours. Marwan would watch them. Abu Laith, who had seen in the Iran–Iraq war what lay at the bottom of those plumes, looked away.

Across Mosul, those who could were stockpiling food, filling basements or cupboards, safe from looters. The animals – despite Marwan and the children's efforts – were still hungry. Rice and leftover chicken bones could only take them so far. The lions needed meat, and the bears needed fruit. It had been a long time since Zombie, Mother and Father ate properly, and it seemed to be affecting them all in different ways. Zombie was growing restless, pacing around his cage. Father was wasting away, just a ragged coat and bones, more dog than lion aside from his great head. Mother, however, was angry. Each day she yowled louder. If Marwan didn't feed her fast enough, she would swipe at him. She mainly ignored Father, though she shared a cage with him, and Zombie, who was in the cage to her left. To her right, in a cage that faced directly on to hers without a gap between them, were Lula the bear and her cub, Warda.

The water was glittering around Marwan's feet when an animal scream made him drop the hose. He turned to the lion's cage and was almost sick. Mother's face was up against the bars of the cage, and she was shaking something very hard. Thick blood pumped on to the ground, and Lula had her paws up against the bars, roaring.

Mother had the young bear Warda by his paw through the bars and was tearing at him, back and forth, making him flop like a fish on land. Lula was trying to fight the lion off her son through the bars of the cage.

Marwan couldn't think straight. 'Stop,' he later remembered shouting, running over to the cage. 'Get off her.'

There was a shovel leaning against one of the cages and he picked it up. The lion and the bear were still screaming, tearing at each other. He took the shovel and stuck it through the bars, taking a blow at the lion's side. She twisted but didn't let the cub go. The blood was pooling on the floor. He took the shovel and hit the lion again, a thwack on the ribs that should have bent her in half. Hissing, she dropped the cub and threw herself at Marwan, smashing against the bars of the cage, her black eyes flat.

Marwan ran over to the bear's cage. Warda was curled up, shaking. Thick, red blood was seeping from the stump of his arm. Lula wailed as she petted her bleeding child.

Mother was roaring at the lost taste of blood. In her paws she held the bear cub's arm, tearing at it with her teeth. Zombie, agitated, paced wildly up and down. Father cowered in the corner of Mother's cage.

'Help,' Marwan shouted over Mother's roaring. 'I need help.'

He could hear the men who operated the rides and maintained the park running towards the enclosure. They needed to get inside the cage, he thought. If they could get Warda out and take him to the vet, they might be able to save him. But Abu Laith had taught him that nothing was worse than getting between a mother and her cub. Lula would kill him before he could take Warda away.

They would have to separate them.

'We need to get her into the house,' he told the others. Inside the enclosure there was a small hut where Lula and Warda slept and were sheltered from the rain. Haji Faez, the ancient caretaker, who trimmed the plants and occasionally swept up by the food stalls, seemed to understand.

'I'll lure her in,' Marwan said. 'Then one of you can pull the gate down behind her.'

There was a gate to the small house that could be closed by someone sitting on the roof of the cage. One of the others jumped up, the bear roaring below him. Marwan ran around the back of the cage with Haji Faez. The house had a small window there, and if they could lure her in from the enclosure, the gate would slam behind her and they could rescue Warda without being attacked.

But hard as they tried, Lula wouldn't leave her cub. Marwan thought hard. He remembered how much she liked apples, and one of the workers was dispatched to the fruit market to fetch some. Haji Faez and Marwan, using the limited tools at their disposal, tried to entice the bear into the house. She was roaring still. Occasionally, she nudged her cub with her nose. He was barely moving except for his chest, which rose and fell.

The apples arrived.

'Throw one over by the hut,' Marwan said.

The apple thudded to the floor, and Lula looked up. She twisted around and made for the door of the house, where the apple lay. Marwan threw in another few from the back window.

'Go on,' he shouted. 'Come here.'

Lula walked slowly through the door. In a rush of metal, the gate slammed behind her. She threw herself against the bars and roared. Separated from her cub, she was squealing in terror. Marwan ran around the front of the outdoor enclosure and opened the door. The cub was writhing on the ground, thick blood smudged all over the floor where he had stumbled in his agony.

Marwan ran to the cub. Gently, he cupped his hand under Warda and turned him over. What remained of his arm was sticky and warm. In the bark-brown fur, there was a deep hotness. As he moved him, the young animal whined.

'Does anyone have a car?' Marwan shouted, lifting the bloody cub. 'We need to take him to a vet.'

Haji Faez did, and in a moment the cub – now in a small cage – was stowed in the back. Two workers drove off, and Marwan looked down to see his hands were covered in Warda's blood.

Abu Laith blamed himself, when he learned what had happened. 'We should have put something between the cages,' he said, as the tears flooded down his face.

Marwan had told him how sometimes, Warda would scamper up to the lions and take a swipe at them through the bars of the cage. But they had both thought Mother just found it annoying. Warda was the size of a beach ball, and it had seemed impossible that the lions might reach through the bars to kill him.

Abu Laith was inconsolable. It was partly his fault, he later said, for trusting Mother. He should have bought another cage, and placed the lion and the bear cub further away from each other.

They waited until the evening, no one saying much, until they heard a car draw up outside the zoo. Marwan ran over to the gate. The caretaker's car had pulled up on the side of the road.

'What happened?'

Haji Faez was carrying the cage in his arms. Marwan's stomach sank. But then he saw that Haji Faez was smiling. 'He's all right,' said the caretaker. 'He'll survive.'

'Warda,' came a shout from behind, and Abu Laith, who had been waiting at the house, crashed towards them. 'He's alive,' he hallooed.

Together they trooped to the bear's cage and put Warda inside, closing the gate behind him. Lula was still wailing and moaning from inside her house. Marwan jumped on to the roof and raised the gate.

The bear bounded out, roaring as she went. Lula was far skinnier than she had ever been, but she was still a terrifying sight. Then she saw her cub on the ground, limping towards her on three paws. She wailed, and wrapped herself around him. Trapping Warda against her stomach, she licked him all over the bandage wound around his stump and over his head.

Marwan took an apple and threw it into the cage. The bear, he thought, deserved it. But for the first time, Lula – hungry as she was – didn't take it. She was with her cub, and nothing could distract her.

Leaving Lula to her son, Abu Laith, beaming, turned around and walked away from the cage. Next to him, the younger children were kicking the bars of the lion's cage, shouting and crying at Mother.

Furious, he grabbed one of them. 'Don't you dare touch that lion,' he shouted, as the children cowered. 'She can't help being an animal.'

The children were less forgiving. That evening, as they sat on the verge by the zoo, they grumbled together. None of them had ever really liked Mother, and Warda was their friend. Abu Laith said that Mother had eaten Warda's arm because that was the law of the jungle, and that that's what lions did. She was extremely hungry, and nothing could be done about it. But the children didn't care. All of them were hungry, too. For weeks now, they had been running and cycling for miles every day to find food for the animals. Mother didn't deserve the food they were giving her.

'We need to do something about Mother,' said one of them. 'She'll kill Father next, and then probably Zombie.'

Abdulrahman spoke up. 'We won't need to do anything,' he said. 'I saw it on National Geographic. Father and Zombie will throw her out of the pack for what she's done, and then she'll pine away and die.'

And the children, thirsty for revenge, nodded sagely.

1. Abu Laith, self-appointed zookeeper, stands by Zombie and Lula's cages

2. Hakam Zarari playing guitar at his home in Mosul

3. Zombie in the zoo
by Abu Laith's house

4. Lula in her cage at the zoo

5. Abu Bakr al-Baghadi, self-proclaimed caliph of the Islamic State, addresses worshipers in the al-Nuri mosque in 2014

6. Mosul Zoo, a few weeks after Isis was pushed back from the area

7. Iraqi armed forces fire artillery at Islamic State targets during the battle for Mosul in 2017

8. Displaced people from Mosul escape the fighting between the Iraqi army and Isis

9. A member of Iraq's Federal Police runs after firing a Rocket Propelled Grenade at Isis positions in western Mosul

10. Dr Amir at the Four Paws office in Vienna

11. A field hospital set up by an Isis sign on the outskirts of western Mosul during the battle to liberate the city

13. Mosul's Old City, which was destroyed during the battle to oust Isis from Mosul

12. A ruined street near a front line in western Mosul

14. Zombie looks on as his mother is buried, soon after Isis fled the area

15. Dr Amir treats Zombie as they prepare to move the animals

16. Hakam and Dr Amir during the rescue mission

27

Imad

2010

SARA'S HUSBAND WAS RELEASED FROM JAIL JUST A FEW short years after Imad had turned up on her doorstep, reeking of whisky and declaring his love for her.

Almost immediately, Imad had started spending a lot of time in Baghdad with his children, where they lived together with Sara and her three kids. It was as if nothing has changed in the twenty years they had been apart.

Three months after Imad had come back, they went to see a cleric for a temporary marriage that would allow them to spend more time together without causing scandal in the family. Though they were Sunni, they had gone to a Shia cleric, who would be less likely to know any of their relatives, and Imad had slipped him a few dinars to turn a blind eye to the fact that Sara was already married.

At the weekends, they would go to the park with their brood and watch as they ran riot across Baghdad's manicured public spaces. Imad would get in trouble for feeding ice cream to the animals in the zoo in al-Adhamiyah. But one day came an announcement that there was to be a prisoner amnesty, and that Sara's husband would be included. From jail, he sent word that Sara had two choices. She could be with Imad, and never see her children again, or stay with him, and see them grow up.

It wasn't really a choice at all. Imad would have to go.

It was a development that, while unpleasant, was accepted as completely inevitable by everyone involved. Neither Imad nor Sara were overly sentimental, and while they loved each other deeply, they accepted that this was the hand they had been dealt. Besides, both knew that they would still be able to see each other, as long as they were subtle about it.

So it wasn't long afterwards that Imad found himself again back in Mosul, alone with his children in his house by the park. The life of a bachelor suited him extremely well. Each day, he would wake with the sunshine that streamed through the window, and the screaming of a dozen small-ish creatures roving about the house.

He had taken quite happily to being everybody's mother, and cooked and shouted and restrained when he needed to do so. The others had become relatively proficient at cleaning and animal husbandry, the two other major skills that were needed in the family.

At breakfast, he would lecture the children on the mating habits of dogs (let them get on with it) and the feeding schedule for ostriches (in the morning, and not too much). When he was done, and they had gone to school, he would go to his mechanic's shop, where he would fix American cars for $300 a job.

By the standards of Mosul, he was a rich man, and proud of it. When he went to dinner with his friends at one of the mezze places in the city, he would always pay, threatening any man who tried to stop him. He wore a long thobe that reached down to his sandals, and grinned over the starched collar, his sienna hair slicked back over his great head.

Despite the drinking and the dogs, he had become a respected man, known for his generosity and his fighting prowess. Since he had built a mosque, he was considered a pillar of the community, and no one really minded that he had stopped praying, and held regular afternoon drinking sessions in his courtyard, whisky and arak spilling on the floor as he laughed and cracked dirty jokes with his friends.

He was regularly invited to coffee at the homes of various neighbours and members of his extended clan. Every time he would go, invited on the pretext of meeting an ancient great aunt or uncle, there would be a shy young woman serving him coffee and making eyes at him. He'd wink at her, but his heart wasn't really in it. They all seemed a bit meek and retiring. He couldn't imagine any of them wrestling with the children, or giggling with him late at night, taking turns to swig from a bottle of whisky.

That made no difference at all to his family, who decided he needed to marry, whether he wanted to or not. Besides, they had a lingering, and entirely accurate, suspicion that Imad had been planning to incapacitate Sara's husband.

It wasn't until Laith, Imad's oldest son, was getting married that anyone took serious action. At the reception in the house, a local sheikh came and – at the prompting of Imad's daughters – sat down next to the man of the house. 'You need to get married,' the sheikh said. 'You have your own daughter-in-law now. There needs to be another woman around to help her in her work.'

Imad, who had heard it all before, said he wasn't interested. He could cook and clean as well as anyone, and didn't need help.

'There's a woman in the Green Flats area,' said the sheikh, ignoring him. 'She's never been married, and she's in her forties. I'll go and see her with my wife and arrange something.'

The idea of going to meet an old spinster did not interest Imad in the slightest but – as his daughters had astutely predicted – he felt he couldn't argue with the sheikh.

A few days later, Imad knocked on the door of a house in the Green Flats, accompanied by the sheikh. They were welcomed in and went through into the living area where the family entertained their guests. Sitting on the floor cushions were an older man and two younger men. Next to them were two girls. One was dressed in a full-length black robe, and her hair and neck were covered in a tight-fitting black scarf. She wasn't smiling, nor was she talking.

The other one, who had leaped up when Imad walked through the door, had already shouted out a greeting and was grinning at him through a pale, round face framed by a long curtain of black hair. Her eyes were rimmed with kohl, and she was wearing a purple dress that clung to the dumpling curves that Imad so admired in women. She was about thirty, while the other girl looked a few years older.

He hoped that it was her he was supposed to find interesting. Grinning, he looked at the sheikh, who looked meaningfully at the other girl. Imad sat down, disappointed.

The men introduced themselves, and took their place on

the cushions along the walls. The girl in the purple dress, who had a high, rapid-fire voice, was already questioning Imad.

'Do you own your house?' she asked, as the family gaped at her. Her sister – the woman Imad was here to meet – looked at the floor mutely.

Imad, who didn't want anyone marrying him for his wealth, lied and told her that he rented his house, and that the only car he owned was a taxi.

'And how is your business?' Imad asked one of her male relatives, politely, as he stared at the loud girl. While he was there, no one would discuss the woman, or the issue of marriage. It was, to all intents and purposes, a simple social call, with enormous implications.

The drab woman he was supposed to be marrying still hadn't said anything. Imad didn't know what to make of her. She seemed very devout, and had completely ignored him. Her sister, on the other hand, was quizzing him so fast he barely had time to answer – scandalizing her relatives with her boldness. There was so much life in her, Imad thought. He knew, without a slightest doubt, that he wanted to marry her.

With another round of hand-clapping and well wishes, they left the house half an hour later. As soon as the door slammed, Imad turned to the sheikh. 'Who was the other girl?' he asked. 'She's the one for me.'

'She's her sister,' the sheikh said. 'I warn you, she's a widow, and already has three children.'

They walked for a minute, Imad cheerfully unconcerned. 'What's her name?' Imad asked.

'Lumia,' the sheikh replied.

Imad was grinning. 'She'll say yes to me.'

28

Hasna

IN THE GREY LIGHT OF THE MORNING, IN THE BIG HOUSE near the Tigris, Hakam's sister Hasna awoke to another unpromising day. It had been months since she had left the house, and even then she had been bundled into a car, every inch of her skin covered, and taken to relatives' homes for a quick visit.

Her exams had finished a long time ago. She had all but stopped going outside when the public executions became more common, and rumours began to thrive about the excesses of the all-female morality police called the al-Khansaa brigade. She'd heard that they pulled over women who were dressed immodestly and beat them with electric cattle prods or a metal tool called the 'biter' that ripped through cloth and skin.

Hasna had taken her course books home with her, thinking that she could at least study with her friends. But then they couldn't leave their houses either, so in the end, she studied alone. Though sometimes she was afraid, she was mostly just bored. The government had disconnected Mosul's broadband internet to stop Isis spreading its propaganda online. The

Zararis now had a satellite connection like everyone else, but they all knew that Isis was monitoring the internet traffic. She no longer dared chat to her friends.

Her computer, where she had once spent evenings on Facebook, stood untouched. Instead, twenty-one and wistful, she lay on her bed reading and re-reading Wuthering Heights. Her copy had a photo of the British actor Tom Hardy on the cover.

She went downstairs through the airy entrance hall. The walls were hung with mirrors and fine-wrought drapes, every surface scattered with hand-knotted cloths and elaborate calligraphy. The skylight let the sun play over the walls.

They had been lucky to keep a house like this. All around them, high-ranking Isis members had taken over the homes of wealthy Moslawis. Most of their neighbours had fled to Kurdish territories or to Baghdad. Others – the ones who lived in even wealthier areas – had been evicted to make space for foreign Daeshis.

They'd had to adapt to survive. Her father Said, a lawyer, had gone from being a towering, stern head of the community to living below the radar – keeping to himself, staying inside as much as possible. He despised the militants with a hate that was almost solid in its intensity, but he couldn't risk leaving his family alone if he were taken.

Arwa, quite as much as Said, bristled at being forced to stay at home. Like her daughter, she was very tall and quick to laugh, with an interrogative lawyer's brain that made her despise unfairness of all kinds. Stuck at home, they had settled into an uncomfortable torpor. At first, they had tried reading all the books in the house. But after they had hosed through Miles Copeland, obscure texts on Islamic jurisprudence and – in the case of Hasna and Hakam, both fluent English speakers – Shakespeare's plays, notably *Twelfth Night*, they had all started to go slightly mad.

What they needed, Arwa and Said had decided, was a hobby. One that didn't require electricity, which had become increasingly intermittent, the internet, or anyone leaving the house. They settled on gardening – specifically the cultivation of rare rose breeds. Each day, the two lawyers crouched in their flower beds, splicing and watering the delicate stems until the garden bloomed. Arwa assiduously documented each stage of growth on her iPad, which she carried around with her in the garden as Daesh roamed the streets nearby. After a few months, Said announced proudly, they had cultivated something extremely unusual. It was a yellow bloom, the edges tinged with pink delicate as ink touched on wet paper. It was a Mosul rose.

Hasna and Hakam, the internet taken from them, became obsessive bird-breeders. Hakam or Said would buy them from the animal market near the house when they were chicks, before they'd grown too close to their parents. The Zarari children would be their mothers instead. Together they would boil and mash chicken eggs and potatoes to a paste, then feed them through syringes repurposed from Hakam's lab. Sometimes he would add some protein powder to the mix, and the birds grew at an alarming rate.

A cage the size of a washing machine stood on the covered veranda. Most of the time it was empty, the door open as the birds flew around the house. At night they slept in Hakam's bed with him, curled up on his chest. When he walked around the house, he would sometimes keep a blue and purple lovebird called Tutu in his shirt pocket. As it grew older, the bird had started eating from the family's plates during dinner, developing a particular taste for biryani.

The cockatiels lived in the kitchen, where they had a cage that almost always stood open. Hakam had raised them to expect to be let outside for a night-time flight around 8 or 9 p.m., and if he forgot to open the front door, they would squawk at him until he did.

One morning around 4 a.m., Hakam was woken by his bleary-eyed father, who had risen for dawn prayers. 'I'm not sure how to say this,' Said had said, unimpeachably calm. 'But your bird is having some issues.'

Half-awake, he'd run downstairs to find Susu, a tiny bird with a feather coat of deepest azure, hysterical in the kitchen. During a night-time exploration mission under the cupboards, she had been caught in a sticky mousetrap. Said had found her hiding behind the air conditioning unit in the kitchen, squawking hysterically. He'd pulled off the trap, but the glue had immobilized the bird on one side.

Still almost asleep, Hakam had grabbed her and, as Said returned to his prayers, proceeded to surgically unstick each feather using a pair of tweezers and a fish knife. The operation had left Susu almost bald on one side, and while the feathers regrew she had been reduced to waddling around the house on foot.

As Hasna went through the kitchen, the cockatiels flew around her head in the morning sunlight. Her father was standing in the living room, looking at the bookshelves with an expression of deep concern.

It had been strange seeing her parents, powerful figures who debated the finer points of Islamic and Iraqi law over the dinner table, turn into reclusive gardeners who barely left the house. Hakam, with his uncontroversial government job, was the only one who wasn't questioned too closely at checkpoints, and was now the only member of the family who left the house every day. Arwa, like Hasna, hadn't been outside the peach walls that surrounded their house for months.

'What are you doing?' Hasna asked her father.

Said took a book down from the dark wooden shelves, which covered one enormous wall of the living room. 'These are Iraqi laws,' he said, weighing the heavy tome in his hand. 'They form the basis for nearly every case I've ever argued.'

Hasna took it and scanned the thin pages.

'I've just come back from the market,' her father said. 'The Daeshis have been raiding houses around here.'

Said was still staring at the bookshelves.

'They're looking for *haram* things,' he said. 'We need to clear the house.'

'We need to burn all the books,' Hasna said, with a prickle of horror.

The family's books were their most prized possessions – a culmination of Said and Arwa's lifelong collector's mania. Rare manuscripts on Iraqi jurisprudence, medieval poetry, travelogues and western novels – bought by Hasna and Hakam – lined those dark wood shelves.

They all had to go.

'Come on,' Said told Hasna. 'Help me get them down.'

'But we can't burn them all,' Hasna blurted out. 'I don't want to. And there are too many.'

Her father started scooping hardback, green leather-bound texts on Iraq's founding laws in armfuls from the shelves.

Hakam and Arwa came into the living room. 'Hakam, you

can leave your work books,' Said said. 'I don't think they'll mind about those. But everything else is going.'

Arwa rushed over to help Said, and together they formed a chain carrying the books from the living room into the garden, which the sun was just beginning to heat. Hasna and Hakam both went upstairs to their rooms.

On Hasna's desk lay *Wuthering Heights* and her copy of *Twelfth Night*. She grabbed them and put them in a shoebox, piling the rest of her books on top of them, which she hid at the top of her wardrobe. Whatever her father said, she was not going to burn them.

She went to Hakam's room. He was standing on a chair, trying to pull out a ceiling panel above the wardrobe. 'What are you doing?' she asked.

The panel came out with a puff of dust. Hakam put it on the floor beside the chair. 'Pass me my guitar and the books,' he said, standing up to peer into the space above. 'I've got to hide them.'

Hasna handed them up, and he stowed the guitar carefully inside the ceiling cavity, putting the books in afterwards and closing the panel behind them.

'I've hidden mine too,' said Hasna.

'You're screwed if they find them,' said Hakam.

'They won't,' said Hasna.

By the time they returned downstairs the living room was almost empty of books – bare shelves marked with their dusty imprints. Through the garden door, Hasna could see an untidy pile of novels and law books lying half-open in the sun.

While Said ran around the kitchen looking for matches, Arwa was doing a last sweep of the house for any forgotten books. The recipe books could stay, she decided, but the novels would have to go.

She took the rest of the books out into the garden. Already the pile had grown to chest height, the silver-embossed writing on the covers glinting in the sun.

Said came out into the garden. No one cared about the law, about doing the right thing, more than him. This was painful. 'Some of them are too valuable,' he said, with an air of forced calm. 'I won't burn them.'

'They can't find them,' said Arwa. 'They'd kill us.'

'I know,' Said said. 'But they won't. We'll bury them deep, and they'll never find them.'

Twenty minutes later, a deep trench was taking shape in the soft ground under the orange trees as Hakam shovelled relentlessly. No one, said Said, would ever bother digging that far on the off chance that something might be hidden.

The great lawyer stood with a pile of books on one side and a bin bag on the other, deciding which to bury, and which to burn.

'I remember the first time I read this,' he said, weighing a hardback tome in his hand. 'I can't have been older than you when I bought it. I was full of ideas.'

Hakam didn't say anything. It felt like his father needed to speak.

'This one taught me about fairness,' Said said, picking up a philosophy tract. 'I haven't read it for years.' He threw it in the bin bag, to be buried. Others – heavily-thumbed works of literature and law – went on the pile for burning. As he threw the books on the bonfire Said seemed to be physically hurting. 'It has come to this,' he said, as he tied up the bin bag. He handed it to Hakam, who laid it in the trench. 'This is what comes of ignorance.'

Hakam shovelled soil back over the bag, sending the books into the ground. The family watched in silence as their belongings disappeared beneath the earth. All Hasna could think of was how stupid it was for Daesh to be so scared of books.

Soon the soil was patted down. Hakam put away the shovel, sweating from the exertion. He took the tank of gasoline that they'd brought out from the house, and poured it in great glugs over the books in the pile. Said struck the first match. As the orange glow spread, the smoke from the burning texts drifted away into the Mosul sun.

29

Abu Laith

THE MAN FROM YEMEN WOULD KILL HIM, MARWAN KNEW, easy as breathing. He had a patchy beard dyed bright yellow with henna, and straggly black hair on his head. His face was very dark, and his body reed-thin, his eyes dull with a dead anger. When he strode into the zoo that day carrying a long black rifle, Marwan had known that there was something wrong with him, and that this wasn't a normal man. Even the other two Daeshis who had come with him said that he was crazy. They were from Tel Afar, a city 40 miles west of Mosul. Next to their leader, the Yemeni, they had seemed reasonable.

'Don't come back tomorrow,' one of them had said to Marwan as the Yemeni walked around the zoo, shouting at the few families who had already arrived to get out. They had left their picnics behind as they scrambled to leave the zoo. 'He'll kill you. He's so angry.'

They had screeched up in a white Nissan van around lunchtime. There hadn't been many people in the zoo. The fighting had been growing closer, the explosions louder, for a few days now and the visitors had dwindled to a few local regulars who took their picnics to the park and came to look at the monkeys. There was no queue in front of the ice cream stand and the lights on the merry-go-round had been turned off. Everyone had worried they would be an easy target for coalition aircraft, and there wasn't in any case enough fuel for the generator.

Marwan had come to the zoo anyway, because he was still getting paid and because he was waiting for Heba. She hadn't been to the zoo for a few weeks, and he had been worried in a way that he had never been before. He missed her, and was very afraid for her. Fighting had been raging around Gogjali, the suburb of Mosul where she lived, for weeks throughout November 2016. He didn't know whether her family had stayed at home, or fled further into the city away from the fighting. He didn't know her address, or her father's name. He had started to become aware that he didn't know whether she was dead or alive. But still he stayed, hoping she would come.

He was also worried about the animals. Lula didn't like loud noises, and the planes were making her panic when they swept through the sky at high speed, making a swooshing, booming sound. She lumbered restlessly around her cage, occasionally pawing at Warda to keep him away from the lions. Marwan had put up chicken wire between the lion and bear cages, but it hadn't stopped Mother staring at the cub as if he was a steak while he scampered about on his three legs, looking surprisingly cheerful apart from his watery eyes.

But now the men from Tel Afar had come to the zoo with the Yemeni, who strode around radiating malice and brandishing his rifle.

'Get out,' the Yemeni had shouted, as the families grabbed their children and ran out of the zoo. 'Don't come back here. This is the property of the *dawla*.'

Marwan hadn't needed more encouragement. The men

from Tel Afar had told Marwan to run, and he ran. He didn't know why they were there, but this was a war, and he wanted to survive. He thought of Heba, and of what would happen if she came to the zoo looking for him, and he wasn't there. The idea, he explained later, made him feel sick with worry. He thought of the animals, of Nusa and the baboon, the lions and the bears. There would be no one left to feed them. They would starve, but he couldn't stay. Marwan's family treated him badly, but he had a responsibility to look after them too. He had to take them to safety, away from the fighting, the only way that was open to them – further inside the city, away from the army and further into Isis territory. The animals would have to manage. He ran out of the zoo gates.

Marwan wasn't the only one who ran. Ahmed, the park manager, had also made himself scarce, Abu Laith realized later that afternoon, after Marwan had come to say goodbye. The manager had worked closely with Daesh to keep the amusement park with its zoo open. According to Marwan, he had almost never been to see the animals, only taken the money at the main entrance. When the army came, he worried that they would punish him for having worked with the militants, so he fled, leaving his charges behind. It just proved, Abu Laith later remembered thinking, what he'd always said: animals were better than people.

That afternoon, Marwan and Abu Laith sat in the courtyard. There was still a last touch of summer in the air.

Abu Laith couldn't stop Marwan from leaving. The war was growing so close that the house sometimes trembled from the mortar strikes, and he knew that the young zookeeper needed to run. But the animals could not survive without him.

'I can't go in there,' said Abu Laith, dejected for once. 'They'll find me. I can't feed them.'

Marwan's body pumped with adrenaline. The Yemeni had terrified him. 'I have to go,' he said.

What worried Abu Laith most was that the zoo was, without question, a legitimate target. There were Daeshis there most afternoons – he had told the army so himself. Whoever was coordinating the strikes on this neighbourhood would hit them first. Zombie would die.

'Where are you going to go?' Abu Laith asked Marwan.

'I'll go further into the city,' said Marwan. The mortars had started up again, screaming nearby. 'I'll see you soon, inshallah,' he said, and ran out of the courtyard.

Abu Laith considered his options. He wasn't particularly worried about Marwan, who was a wily sort, and would probably be fine, despite his broken heart. But the realization that the animals would die had hit him hard. He couldn't see a way around it. The city had never been more dangerous. The war aside, the Daeshis were hunting everywhere for spies. Marwan, during their debriefs, had told him Daesh was becoming more paranoid even than before. Rumours had spread that everyone with government connections would be arrested. In the centre of the city, dozens had been beheaded, and videos of the executions posted online.

Many of those killed were former government employees – or, like Abu Laith, former soldiers – who had only stayed because they had trusted the Daeshis, who told them it would be safe. When the militants had first taken power, all former members of the security forces had been forced to sign a *towba*, a confession of past wrongdoing and a pledge of allegiance to the Islamic State. The Daeshis had told them that it would absolve them of their crimes.

Many had sensed a trick, and stayed in hiding. They had been right: now Isis was sending death squads to those who had trusted them – dragging them from their houses or from the streets or into their prisons, cutting their heads off or shooting them.

Food was even more scarce than before. The streets of the Old City were clogged with Isis supporters displaced from areas retaken by the army, civilians the militants had taken as human shields, and others who had remained living there.

For Abu Laith, leaving was not an option. Marwan could flee with his family further into the city, where they might be safer – at least until the army came. But Abu Laith was a wanted man. He would have to stay, and the family with him, until the new invaders with their American guns and vehicles swept back into the neighbourhood.

They needed a plan, Abu Laith decided.

The next morning Luay, the most inconspicuous of the family with his brown hair and long beard, left the house on his father's orders.

'Sneak up on them,' Abu Laith had said, as he stood by the gate to see him off on what he intended to be a very short walk. 'We need to know what's happening.'

Affecting a nonchalant air that he did not feel, Luay checked the road and walked up to the zoo fence. He glanced into the park. If the Daeshis caught him looking, they might think he was a spy, which he was, in a sense.

As he walked up the road, he saw them. Three Daeshis were sitting with their backs next to a wall on the other side of the zoo. They were wearing beige Kandahari shirts, and in front of them was a heavy mortar. He saw others scattered around the grounds.

Luay forced himself to turn around slowly and walk back to the house.

'They're doing what?' shouted Abu Laith.

'They're moving into the zoo,' repeated Luay. 'I guess they're using it as a base.'

Abu Laith sat on his favourite sofa and fumed. But however much he thought, he couldn't find a way out.

The roar of vehicles stopping outside snapped their heads to the door. Abu Laith ran to the stairs. They waited, but no one came to the door. There was a mighty clattering outside, punctuated with shouts and whelps from the fighters as they clanked about with what sounded like very heavy objects.

Abu Laith slunk up to the top floor, on to the roof. He looked down into the park. They were, as Luay had warned, turning the zoo, with its starving animals, into a Daeshi fighting position. There was a gun emplacement by the pelican pond. Black sticks, instantly recognizable to an old army hand as mortar tubes, were stacked up by Lula's cage. Everywhere the fighters moved like armed ants.

By that afternoon, it was finished. Lula and the lions were ringed by mortars. Everywhere gruff men with strange accents bustled about, carrying shells or shouting orders.

They were, Abu Laith reflected, very efficient.

30

Abu Laith

The sun rose like an egg cracked in a pan, its warm yolk steadying the new sky. On the roof, the pigeons were still sleeping in their cots of sawdust and excrement. On the stairs, Lumia was dragging up a sack of wood, making a considerable racket.

'Abu Laith,' she shouted. 'Abu Laith.'

'I'm coming,' barked Abu Laith, who was carrying another, much larger, sack of wood. 'Keep going.'

The children scampered around him, getting under his feet and bombarding him with questions. Abu Laith barged up the stairs after Lumia, a cloud of dust rising in his wake. 'We're going up to the roof,' he shouted at the children over the din. 'We're baking up there.'

Lumia had stopped right at the top of the stairs by the door that led to the roof. She put her bag of wood on the floor and peeked through the cracks. All was sun and golden light, the first touches of winter. She listened. There were no aircraft in the sky. 'Can we go?' she asked, as Abu Laith crowded up next to her on the stairs.

He burst through the door on to the roof. Lumia forced

herself to calm down. 'Children,' she shouted down at the crowd gathering on the stairs. 'Stay on the stairs. If a plane comes, I want us all to die together.'

None of the children minded. They were used to fatalistic proclamations before breakfast, and were very much looking forward to seeing the bread-baking.

The war had drawn much closer since the afternoon Marwan had left. For days, the house had shuddered with the jolts of mortar and artillery shells. Sometimes they heard screams, and the pop-pop of rifle fire.

Abu Laith had been kicked into action. The front gate had been locked with heavy chains, and everyone banned from the garden. The windows were left open so they wouldn't shatter in a blast wave. Water, drawn from the well in the mosque courtyard and fed to the house by a hose that had been there for years as a back-up water supply, stood in buckets by the main bedroom. The children had made their beds next door in the second living room, dragging down mattresses to form a spongy floor several inches thick where they rolled around and played all day, or lay at night, piled up against their parents.

It was the room furthest from the road, where the fighting was likely to be, and it had sturdy walls that might withstand a hit to the roof. To the back, it was protected by their neighbour's house.

When they did come, the planes rushed over the house so fast the children cowered like hunted rabbits. The mortars weren't as scary, they thought, but Abu Laith knew better. They landed almost randomly and vertically, fired by the army from miles away into neighbourhoods where people lived tightly as notes in a wallet. A direct hit could penetrate the roof. As it was, one had already hit the house behind them, missing their neighbours but burrowing its way through the walls between the houses. It had shattered the back window and sent Lumia running through the living room screaming to check on her children, none of whom could hear her as they had all gone temporarily deaf from the blast.

That afternoon, when they could hear again, they had tidied up the dust and the glass, and Lumia had cried with the smaller children, while the older ones inspected the damage to the back wall of the house. Abu Laith had told Lumia she could try to flee if she wanted to, go further into the city like all their neighbours had done, but she had just shouted at him, and he hadn't said anything more about it.

Then there was the matter of the Daeshis in the zoo. They could decide at any point to use Abu Laith's house as a base, or his rooftop as a gun emplacement or a sniper position. Then they would be a legitimate target for the American aircraft which – they had heard – could kill a single person on the street, or destroy a whole apartment block, just by a man pressing a button miles away.

Food was the biggest problem. Abu Laith and Lumia had stockpiled rice in the kitchen, along with beans and tomato paste. But they hadn't had enough time or money to buy cans of food, or onions, or anything that would do more than let them survive. As it was, Lumia still needed to go into the kitchen every day to cook. It was at the front of the house, facing on to the garden and the road. Each time she went in, she took a moment to prepare to die.

A few mornings ago, Abdulrahman had told her the window was broken in her and Abu Laith's bedroom. She had rushed in, but nothing seemed to be wrong, until she found the newly chipped wood on the bedframe – right where Abu Laith's head usually rested – that showed where the bullet had struck. It had passed over both of their pillows on its way into the room.

As it was, there were no other options than to get on with it, so she shut the wailing children into the bedroom and crept into the kitchen each morning, hunkering between the counters as she filled one of her metal pots and set it on the stove. They still had a little gas. She would light it and hide behind one of the counters until she heard the patter of the boiling water.

But it wasn't enough.

'We need bread,' she told Abu Laith a few days after the war proper had begun. 'We won't survive without it.'

Usually Lumia bought flatbread – without which no Iraqi meal is complete, rather like baguette in France – from the bakery a few streets away. But now a walk there was tantamount to a death sentence, and the bakery was almost certainly closed anyway. It was a good thing that Lumia had been through some hard times before she married Abu Laith, a relatively rich man, and knew how to make bread on a stove.

'Come on,' Abu Laith said to Lumia, as she peeked her head out of the door to the roof. 'It's clear. Come on.'

With a deep breath, Lumia did one of the bravest things she had ever done and stepped on to the roof, crouching so that they wouldn't be seen from the zoo. There were still Daeshis down there, firing mortars at the soldiers who were far behind the house somewhere.

Lumia put the board with the dough on it in front of her on the ground. She started kneading it automatically, pulling off pieces the size of a child's fist to push into the rough shape of flatbreads, spinning them between her hands to make them thin and light.

Abu Laith was standing crouched over the stove that they had improvised on the roof. He struck a match, puffed to encourage the flame, then fed in the small logs that they had carried up. As the flames began to take, he grabbed a bit of cardboard and wafted the fire with great energy. They could not afford to make smoke – it would be a beacon for the Daeshis in the zoo and for the aircraft overhead. If Daesh knew there was anyone on the roof, they might come for them: to loot and to arrest suspected spies.

'Come on,' said Abu Laith, in what was clearly meant to be an encouraging tone of voice. 'It's hot. It's going to be all right.'

Lumia crawled over to Abu Laith. She took one of the thin circles of dough, dusting her fingers, and spread it over the metal hot plate. Abu Laith was still fanning, standing half-crouched over her. She started flattening a new disc of dough, and Abu Laith flipped over the bread. A moment later, he whipped it off and put the new one on.

Then the air above their heads began to zip and crackle.

'Get down,' and Abu Laith pushed Lumia to the ground. 'They're shooting.'

Lumia's fingers burned on the stove. Abu Laith stayed half-crouched on the floor, scanning the parapet. 'It's OK,' he said. 'It's going over our heads. I don't know what they're shooting at.'

'It's not OK,' said Lumia. 'They're shooting at us.'

'They're not,' said Abu Laith, reasonably. 'Or we'd be dead.'

There was a rattle from the zoo, and another vicious flight of bullets, this time over to the side.

'We have to go back down,' said Lumia.

Abu Laith looked down at her. 'Come on,' he said. 'Finish the bread. We need to eat.'

Lumia, summoning the last of her courage, flattened another ball of dough as the bullets whipped past over her head.

By that evening the children were, for once, quiet – lying scattered in groups across the mattresses playing with their surviving toys. The generator was turned off, and it was dark. When Luay had complained, Abu Laith had shouted at him that if he wanted to go and risk his life to get more fuel and turn the generator on so that he could watch TV he was more than welcome. Luay had declined, and instead lay moodily on a mattress just outside the main room where the rest of the family slept.

As they waited for the army to come, Abu Laith skipped in and out of the back room, up and down the stairs, keeping an eye on the zoo from the roof.

He was just settling down into the sofa in the living room by the back of the house that evening when he heard the unmistakable snap of a bullet. The wall behind him smacked into his head as he threw himself back before he'd even thought about it – reflexes he'd learned in pointless months eating mud up by the Iranian border. Behind him, Lumia screamed.

'Are you hurt?'

Abu Laith sat up, grinning. In the place not far from where his head had been just a moment before, on the left side of the living room doorframe, a finger-sized burrow showed the path of the bullet.

Abu Laith pointed at the hole in the hallway wall. 'It must have ricocheted through the living room, then the door, then through that wall. What the hell can they have been shooting at? The army is miles away.'

'Shut up and get down,' shouted Lumia, who didn't care about this at all, and just wanted her husband not to die.

Still marvelling at the tenacity of the bullet, Abu Laith strolled back to the bedroom, where everyone was cowering. The assault on Daesh seemed to have been renewed.

'There's outgoing artillery from the army, now,' said Abu Laith. 'They'll be here soon.'

Lumia, who didn't understand how they were all being so calm, started to cry again, occasionally wiping snot from the toddlers, who were also weeping.

Nine-year-old Abdulrahman sat quietly. The fighting had scared him at first, but now he was getting used to it. Like all children, he did not seriously entertain the idea that he was going to die. He didn't like the screech of the planes overhead, but the mortars had become more or less normal over the last few days. He was bored.

'Shall we go and look?' he asked Abu Laith, in a low voice so Lumia couldn't hear. The two of them had been sneaking up to the window on the first floor to see what was happening down by the zoo. Last time, they had run back down quickly. The area by the lion's enclosure had been full of Daeshis. If they saw them in the window, there was every chance they would fire at the house. But if they were careful, they could look without being seen.

Abu Laith agreed. Together, they crept out the room and up the dusty stairwell, covered in bits of plaster and cement knocked loose by the shock of nearby explosions. The hallway on the first floor was empty except for a few carpets, and the air was thick with dust.

They crossed the room on their knees, keeping out of sight of the window. As he crawled, Abdulrahman felt his bravery seep out.

'Over there,' whispered his step-father, pointing to the left out the window.

Abdulrahman craned round to look at the near side of the zoo. Just over the wall, by the mosque, men in dark green clothes were moving around. He was very scared, he later remembered.

'Come on,' said Abu Laith. Slowly, they crept back over the floor.

Abu Laith was overjoyed that the zoo hadn't been bombed. Sitting contented in the bedroom, he continually told everyone who would listen that it meant the animals were still alive. With the door to the front garden shut, the bangs were a little quieter. But they still rocked the house on its foundations whenever they fell nearby.

Abdulrahman tried to sleep, but every time he closed his eyes the sound of the rockets and mortars started again in his head, mixing with the real ones outside. Instead, he listened to his father, who was telling Geggo how birds flew.

'Their bones are hollow,' he roared, unperturbed by the thuds around them. 'And their feathers are balanced so perfectly that they slide on the air.'

With bombs real and imagined popping in his head, Abdulrahman drifted to sleep.

The army was still far away.

31

Hakam

The bathroom by the entrance hall had never been a particularly notable feature of the Zararis' home. It was white and clinker-tiled, with a toilet in one corner, a sink by the wall and a washing machine with a boiler above it. The other bathrooms were decidedly nicer.

Now, it was the family's only home. For the last two weeks, they had been living in the bathroom as the airstrikes and mortar shells pounded the neighbourhood throughout the beginning of December 2016. The house shook with the relentless bombardment. Planes screeched above. One night, Hakam had been watching a film on his laptop – headphones in – when there was a great sucking boom that set the house shaking. For about two seconds, he thought they had been hit, and lay there in the absolute knowledge that there was nothing he could do to stop his death.

But it had hit the Daesh 'media point' in the street behind them, and though everyone screamed and held each other and Hakam slammed the laptop shut they knew they had been lucky.

They'd spent the rest of the night awake, lying on their

mattresses on the bathroom floor. Hakam brought an electricity convertor, which he charged from their tiny generator, and they could watch films and listen to music if the bombing wasn't too bad. Said had chosen the bathroom because it had only one small window, to the back of the house, and strong walls that might survive a mortar hit to the floors above. It would protect them if the house next door was bombed, but if an airstrike came too close, they would be dead, the clinker-tiled walls collapsing on them.

At night, the family tessellated into the tiny room, which was barely big enough for the four of them. Hasna and Said slept in the far corner, along the wall towards the sink. Hakam and Arwa slept perpendicular to them, by their feet.

They had known this time would come. As soon as they heard the Iraqi advance had started, Hakam and Said had driven straight to the supermarket. It was already crowded with harried-looking men sent out to ensure their families wouldn't starve.

In the shops, no one spoke to each other. No one argued, even though the shelves were sparse and the prices had risen to absurd levels. The last thing anyone wanted to do was stick out or cause an argument.

The car had been weighed down with food: tins of tuna and tomatoes, bags of rice and beans, oil and flour. Water, gallons of it, in case something happened to their well.

There was barely anyone on the streets, and a lot of the Daeshis looked to be no older than teenagers.

'Take everything to the kitchen,' Said had instructed when they got home. 'Soon it'll be too dangerous to go anywhere else.'

They filled a car with supplies – tins, dates, water, dried

bread – and drove it to Hakam's grandmother's house on the far outskirts of the city. The extended family would meet there if they were forced to leave their homes.

A few days later, the first airstrike hit the neighbourhood. It wasn't close, but from then on, each night was filled with the clapping boom and thud of explosions.

Covering the floor in polystyrene and then mattresses to limit the cold and the damp, the family moved into the bathroom. At first, Hasna and Hakam were scared all the time. But after a few days, boredom set in, and they snapped at each other during the few hours they were awake. Most of the time they lay on their mattresses, half-asleep and uncomfortably hot as the birds tweeted in their cage by the sink. When they had some battery life, they played games on their phones, or watched films – quietly – on their laptops. Everyone had stopped reading.

At mealtimes, when the bombing wasn't too bad, they'd rush into the kitchen and grab what food they could – usually dried bread that they'd soften with water. When it was very calm they could cook, or sometimes even do the dishes. But other nights the bombing was so bad they couldn't leave the bathroom, and just lay there, hoping.

Hassan, in his dorm room at Penn State, suffered with them. Occasionally, if there were no bombs, Hakam and Hasna would sneak up to the roof and text him. 'We've buried the gold in the garden,' they wrote one night, crouching in the dark. 'If we die, that's where it is.'

A few times, the family had woken to the sound of shouting as their neighbours were driven from their homes by Daeshis who wanted to requisition the properties to use as mortar placements or a place to hide sleeper cells.

So far, the Zararis had escaped. But one day around lunchtime they heard loud banging on the gate and the revving of a very large engine. Hakam had been about to undress to get into the shower, but ran downstairs to where his father was striding into the courtyard.

When they reached the garden, Said pushed Hakam back towards the house. 'Hide everything,' he whispered. 'I'll go and see what they want.'

Hakam obeyed, and Said went to the gate. 'Assalamu aleikum,' he said in his best courtroom tones.

'Wa aleikum assalam,' said a gruff, Iraqi voice. 'We are from the *dawla*. Open the door.'

'One second,' said Said. 'It's locked. I'm just going to find the keys.'

He ran through the house and told Hakam what had happened. Together, they went into the kitchen where Arwa was making her famous *orouq,* thin-pounded mince, delicately seasoned, encased in a crisp layer of bulgur.

'Daesh are here,' Said told her. 'We have to go. We have to run.'

Arwa's eyes were still locked on the stove. 'Just wait,' she said, harried. 'I've got to take this with me. It's almost done.' She started to stuff onions into a bag.

'Mum,' Hakam shouted, almost jumping on the spot. 'There are Daeshis who are about to come into the house. You need to get your stuff. We need to leave now.'

Arwa seemed to snap into action. 'Grab your things,' she said. 'I'll tell Hassan we're leaving.'

As his mother ran to get the phone out of its hiding place, Hakam sprinted upstairs. He needed to bring clothes, he supposed, and his phone. Taking the stairs three at a time, he ran into his room. His guitar was propped in the corner. There was no time to hide it in the ceiling cavity. He threw his prayer mat over it, and hoped they wouldn't look too closely. Grabbing his laptop, he locked the room and ran downstairs to hide the computer under the mattresses in the bathroom along with his cigarettes.

Hasna bustled past him, carrying a small bag and half-dressed in her black robes. 'We need to go,' she said. 'Now.'

Hakam still had one, vitally important, thing left to do. He ran back up the stairs and on to the roof, crouching low so the Daeshis wouldn't see him. The satellite dish was hidden behind an old water tank. They had been using it to watch the news as liberation drew closer. But if Daesh found it, they would almost certainly be executed. Hakam pulled it out and ran downstairs, shoving it into a kitchen cupboard and padding it with sacks of flour and wheat.

For all that they had prepared, the family were barely dressed when the Daeshis knocked on the door again.

Said walked into the courtyard and opened the gate.

Outside were four men, all carrying Kalashnikovs and dressed in dusty black and beige. Three were young, dark-skinned fighters with long hair and wispy chins. One was far older: his long white beard reaching down his chest. He was the one who had spoken.

The four of them were standing by the courtyard door. On the driveway, by the garage, was a suicide car covered in homemade armoured plating.

'Welcome,' said Said, pleasantly. 'How may I assist you?'

'Thank you,' said the older. 'Please, it is dangerous here. For your own safety, we will need you and your family to leave. The *dawla* needs this house.'

Said tried to head them off. 'It is written in the sacred texts that Muslims should not be displaced from their homes,' he said. 'As the Prophet, peace be upon him–'

'We're not throwing you out of your house,' interrupted the older man. 'It's for your own safety. The army are coming and they will kill you. You can stay at the house, but it's better to leave.'

Two of the younger men were holding machine-gun parts. Said dropped his smiling, scholarly demeanour. 'Fine,' he snapped. 'Can we take the car?'

The older man considered him for a moment. 'Fine,' he said. 'Now open that door and let us in.'

A moment later the family tumbled through the courtyard door on to the street, the women swathed in black. As they ran to the car, Hakam realized he had forgotten to bring any extra clothes. The kitchen was still full of the supplies they had bought, the *orouq* still on the stove. All they had with them was a heater, a few clothes and some gasoline. The Daeshis were already driving their armoured suicide car into the garage, where it would be hidden from the American drones.

Hakam pulled open the car door for Hasna, jumping in after her. Arwa and Said sat in the front. As they drove down the abandoned streets towards Hakam's grandmother's house – avoiding the main roads, where the airstrikes were concentrated – they scanned their surroundings, every passer-by suspicious.

'We had to go,' Hakam said. 'It was the only thing we could have done. We'll be back soon.'

He hoped it was true.

32

Abu Laith

Two weeks would have been a long time for anyone to be stuck inside a small room, under heavy bombardment, with all the members of their family. But for Abu Laith and Lumia's children, it was torture. Robbed of their usual pastimes of running at full pelt through the neighbourhood, they boiled into an insane, rolling mass of activity. The living room at the back of the house seemed to be full of brown limbs, fighting, poking, stretching, arm-wrestling. Lumia had stopped praying for her family to be saved, deciding it was hopeless, and was instead engaged full-time in shouting at her children to be quiet.

Baby Shuja was already proving himself a true member of the clan by screaming, non-stop, at full volume from inside his cot.

Above the racket, Abu Laith's voice boomed out relentlessly, offering unsolicited advice and opinions on the condition of the animals at the zoo. 'Zombie is alive,' he said, often. 'I know he is.'

It was wishful thinking. None of the animals had been fed or watered for over two weeks. The deafening mortar fire continued from the zoo, and the incoming artillery was just as loud. From a documentary on the National Geographic channel, Abu Laith had gathered that desert animals could live a very long time without water, which gave him some hope.

'Two months for camels,' he had told Abdulrahman, who was extremely worried about the lions. 'It'll be a long time for Zombie too.'

Lumia wondered, rather sniffily, how long humans might last. There was a well in the in the mosque courtyard. Before the fighting started, they had run a hose from it to fill a water tank in the garden that was connected to the taps. But once that ran out, they couldn't risk re-filling the tank. For two weeks, they had rationed hard and eaten little more than beans and rice from sacks that were stacked in the corner of the living room. Without water, they wouldn't be able to cook them either.

Their TV, hidden in a cupboard in the living room, and played only at a whisper, was tuned to an Iraqi news channel. They were far enough from the centre of the city that they could still get a satellite signal, which had been impossible in the centre of Mosul for years. Their satellite was hidden on the roof, and no Daeshi had ever climbed up there to look for it.

But the more they watched the news, the less they seemed to learn. The anchor usually spoke over pictures of Mosul's outskirts, which were in ruins from the fighting.

The news had run the same story for weeks now. The army were closing in on Mosul, taking the outskirts, moving forward slowly but consistently. It didn't seem like it from where the family were sitting in their virtual prison that winter of 2016. The sound of the battle had seemed very close for a long time from behind the house, and Daesh was still fighting in front of them in the zoo. Each day they heard mortars go out with a whoosh, and the crackle of gunfire. Abu Laith could not see who they were firing at. The army seemed much too far away.

Lumia had given up making bread on the roof, and moved her oven next door, where she baked together with other neighbourhood women. When the fighting had started, they had all been terrified, but now they climbed over the garden walls to gossip with each other whenever it was quiet. One day, when Abu Laith had switched off the news, she gathered her baking things.

'All of you,' she said to the children, in a tone that brooked no disagreement. 'Stay inside. Or else.' She looked around for Abu Laith, who was in the living room, playing with Geggo. 'Look after the kids, I'm just going next door,' she said.

'Fine,' he called out. 'I just need to go across the road and clean up the glass. I think the mosque windows have probably smashed.'

Lumia paused in the doorway. 'Why do you need to do that? What's the point? Just leave it. You'll get shot if you go there.'

But Abu Laith wasn't listening, and she couldn't face arguing. Her husband was prone to strange impulses, but he forgot them often as not.

Next door, she lit the fire under her stove and struck up a discussion with one of the neighbourhood wives. As they flattened and flipped the bread, she could hear the children shouting nearby, as was their wont. Occasionally, a bullet snapped. Once, she would have jumped at every sound, but now she was getting used to it, and kept cooking. By mid-morning, when she was almost done with the bread, she saw Abu Laith sneak across the road into the mosque courtyard.

Geggo was following him. 'What the hell is he doing?' she shouted at no one in particular.

In the courtyard of her home, Nour, Lumia's oldest daughter by Abu Laith, watched as her brother and father ran across the road to the mosque. She hadn't been outside for a long time, and was very bored. Before Daesh had come, Nour – now six years old – had worn tracksuits and played hopscotch with her friends in the street. But she had barely been allowed out of the house for two and a half years. When she did go out, she had to wear a long cloak and cover her hair. She had to be quiet, and couldn't gesticulate with her chubby arms and shout the way she usually did when communicating with just about anyone. So when her father and her brother – who was younger than her and had long hair like a girl – walked out of the courtyard door to go over to the mosque, she wanted to go with them.

'Fine,' Abu Laith had said, when she asked him as he was walking out of the gate. 'Bring us a broom. We need to sweep up the glass.'

Beside herself with excitement, Nour grabbed the broom and ran over to the big metal front door. She pulled it open. There was a weird fierce buzzing, then something hit her leg. She looked down and screamed.

Next door, Lumia's head snapped up from the stove. 'Nour has been shot,' someone shouted, she didn't know who. 'Nour has been shot.'

Lumia moved faster than she ever had before, vaulting through the room and into the road with no thought to her own safety. There was blood on the floor by the gate, and she ran through into the courtyard. The children were crowding into the kitchen, and she pushed past them. Everyone was screaming, most of all Nour, who was lying on the kitchen table with her trousers pulled down. One leg was completely red, blood running through her fingers as she clutched at the wound.

'Nour,' Lumia screamed, sweeping her up in her arms. 'What have they done?'

Nour was choking with tears, and kept her hands clamped over her thigh. There was a lot of blood, and Lumia knew she had to stop it, but couldn't think clearly.

'What happened?' shouted Abu Laith, bounding through the door. He pushed past the children and took Nour's leg in his hands. He sighed, and clamped a cloth over it. 'It just nicked her skin,' he said, in a tone of overwhelming relief. 'Praise be to God. She's OK.'

If Lumia had had a gun, at that moment, there was a high possibility she would have used it. 'She's OK?' she screamed at Abu Laith, who was still holding the cloth tight over Nour's leg. 'She's OK? She's been shot, because you couldn't be bothered to look after her.'

Abu Laith, who had not considered this, glowed with embarrassment and anger. 'You can't say—' he began.

'Yes I can,' shouted Lumia. 'You almost got her killed, and it would have been all your fault. She was coming out to help you, when you should have been inside looking after them.'

Abu Laith stared at her.

'If it wasn't for all these children I would divorce you right now,' she said, and meant it.

Outside, there was more gunfire, and the children scattered to the back room. Then there was a thud, a loud one, close, and the kitchen filled with dust. Choking, Abu Laith picked Nour up and the family ran into the back room.

The dust had reached into the house by the time they slammed the door shut behind them, and the air was thick and yellow under the strip-lights. Nour was still screaming, and Abu Laith put her down on the floor-beds.

'It's only a mortar,' he said calmly, unwrapping her leg and washing it. He tied the cloth around it, tight enough so that it would stop bleeding. 'It won't hurt you. It landed far away.'

Behind him, Lumia was trying to make the children get down. Once Nour's leg was tied up, Abu Laith went out into the corridor. The road outside had been hit, he was sure. That morning, another mortar had hit just a few hundred feet away. It didn't bode well. The army gunners were probably aiming for the Daeshis at the zoo. They were getting close. The next one might hit true. Abu Laith struggled to listen, but he couldn't hear the lions' roar over the din. The children and Lumia were still shouting. Eyes blinking against the dust, he walked up to the front door of the house, which faced on to the courtyard.

Then everything was dust and light, mud and sand spraying in his face. He stumbled, and he heard the sound, the crashing boom of an explosion whose pressure made his ears pop and his eyes sting. He looked down, and saw that his body was still all there. When he squinted, he could see the garden wall, intact. Someone was screaming behind him. No lions roared.

'They've hit the zoo,' he shouted, running out into the courtyard. 'It's an airstrike.'

Abu Laith didn't know quite how, but he was in the road, and the gate was open behind him, and the shooting had stopped. He was running at full pelt towards the zoo gate, and the dust was dry and hot in his lungs. He didn't think about the risk, the airstrikes or the mortars or the Daeshis he knew were in that zoo. But someone had killed his lion and he wanted to be near Zombie at the end.

From behind by the houses, he could hear his neighbour, Abu Issa – a deeply serious man. 'Stop, Abu Laith,' he cried. 'Don't go in there. Stop. They'll kill you. The planes are going to come again.'

Abu Laith slowed down, but he didn't stop. He stole around the side of the mosque by the main gates and ran through them, keeping low to the ground. The dust was still thick, and it was hard to see more than a couple of feet away. All over the ground were deep craters the height of a man. The grass lay in tufts crowning clumps of dirt, thrown any old way over the ground. Ahead, by the amusement park, he could see the edge of an enormous hole where the bomb must have struck.

Shapes were moving in the dust to the left, towards the animal enclosure. He realized they were running towards him, feet beating on the ground, and assumed they were Daeshis. Throwing himself aside, he saw three peacocks run past at full speed. Coughing, he jogged closer to the animal enclosure. He couldn't see it through the dust and the splintered trees. He made it to the chicken cage right on the outside, and felt his stomach turn. The chickens were all dead or dying, croaking out their last at the bottom of the cage, red smears all over the floor. The stream by the pelican pool was thick with blood.

He didn't see the first Daeshi until he was almost on top of him – his body splayed out in the middle of the path between the trees. Dusty long black hair was stuck to the ground with his own blood. He was dead. Over to the side, Abu Laith could see another one, crumpled into a heap. Ahead, when he squinted, he could just make out the shape of another, dressed in a long Kandahari shirt and trousers, pinned halfway up a tree, with something long and sharp protruding from his chest.

For a moment, Abu Laith stood still, staring at the man hanging dead from the tree. He needed to get to Zombie. But slowly, his old soldier's brain kicked in. Three Daeshis were dead. But there were always at least four of them in the zoo. If one of them was alive and saw him, he would be killed on the spot. If others came to check out the site, he was dead, too. Above the zoo was a huge black and yellow cloud of dust and smoke. It had to be visible for miles away.

The blood receded from his ears, and he could hear voices behind him. 'Abu Laith,' his neighbour was shouting. 'Abu Laith, come back here.'

Something clicked into place in Abu Laith's mind. He turned around and bolted back to the road through the gates, swinging left past Abu Issa, who was still standing in his garden yelling, and in through the courtyard gate of his own home. Ignoring Lumia's enquiries, he tore up the stairs and on to the roof.

The air was still filled with dust, and great hunks of mud lay smashed into pieces all over the rooftop. Glass and metal turned deadly shrapnel by the explosion lay scattered everywhere. Panting like a water buffalo, Abu Laith crouched down and crawled over to the edge of the parapet. From here, maybe, he would be able to see the animals. Ears tuned for the sound of a lion roaring, he stuck his head above the wall. It looked like the airstrike had hit just down by the carousel.

A muddy hole spread across the lawn where the mortar battery had stood. Nothing was left of it except for a few black spikes here and there. Abu Laith could see the bodies of the three Daeshis – two on the ground, the other pinned to the tree. The whole park was covered in dust, and looked yellow.

He couldn't see much of the animal enclosure, which was half-hidden by trees, but with a pricking sense of relief he saw that the cages were, at least, still standing. The airstrike had landed about 50 feet from them. It would have filled the air with shrapnel, which was surely what had killed the Daeshis. But it was possible that some of the animals might have survived.

Through the haze came the roar of an engine, and a small dust cloud careered down the street towards the zoo. Abu Laith ducked down behind the parapet and peered out through the gaps in the wall. A pick-up truck was weaving through the rubble strewn on the road up to the main gates. He watched it stop by the zoo entrance. A couple of Daeshis jumped out, guns levelled, and ran into the zoo.

Fear rose in Abu Laith, unbidden, that they would kill the animals if they weren't dead already. He watched them run up to the bodies of their comrades. Together, a few of them carried the first two to the pick-up, and threw them in the back. The rest went to the tree, where the third man was pinned. They took him down, and hauled him into the back of their truck with the other corpses. For a while, they hunted around the zoo – looking for the fourth, Abu Laith knew, though he was still stiff with fear of what they could do to the animals.

Within five minutes, the Daeshis were back in the truck. They careered out of the zoo, reversed into thc road and swept away again. Abu Laith took stock. There was only one thing to do, he knew.

He had to get back into that zoo.

33

Abu Laith

'Abu Laith,' screeched Lumia, in a voice penetrating as a dog whistle, and even harder to ignore. 'Abu Laith. What are you doing?'

Her husband, who had one leg over the garden wall on the way into Abu Issa's house, froze in the act of escape. 'I'm fetching food,' he said, choosing his words carefully. Lumia, fear of bullets forgotten, hustled into the garden.

'For who?' she asked. 'For us?'

Abu Laith made an attempt to start a sentence.

'Or is it for the animals?' she shouted, getting into her stride. 'The animals in the zoo who are probably dead, rather than for your children, who are alive and starving?'

The children in question, who wanted nothing more than for the animals to live, crowded around her, jabbering. 'We don't mind,' Abdulrahman said. 'We've found food for them before.'

Lumia tried to usher them into the house. 'You shouldn't be out here,' she scolded. 'Get in.'

Abu Laith took advantage of the momentary confusion to make his exit over the wall. In his hand, hidden from Lumia, were some plastic bags that he hoped to fill before night time. The escaped peacocks had given him some hope. They were the ones who had been closest to the crash site. If they had made it, others might have too. He would go back into the zoo at night, and feed the survivors if he could. They hadn't eaten for more than two weeks. He hoped, with hopeless desperation, that they were still alive.

Abu Issa was standing in his courtyard, where he had been shouting for Abu Laith to come back from the zoo just a few hours before.

'I'm going back in,' Abu Laith said, after the necessary pleasantries had been exchanged. 'Do you have any food I could take them?'

Abu Issa considered, for a moment, his neighbour. He was mad, but he was a good man. Abu Issa went inside and brought out a plastic bag with a raw chicken inside. 'Good luck,' he said, as his neighbour grinned, thanked him and bolted over the wall to the next house with a startling agility.

By that afternoon, Abu Laith had assembled a feast in the courtyard. There was a bag of rice and beans from one house, some half-decayed meat from another. People had been generous. The electricity had been out for days now, and meat was starting to turn. He had taken chicken legs from their own fridge, now relegated to a temperate plastic cupboard, and when Lumia complained he had told her to shut up, which she hadn't.

'Why are you doing this?' she asked, as he piled the food into a big white flour sack. It was half full. 'It's so dangerous. You're going to be killed.'

'Don't you think we should be showing mercy to the animals?' Abu Laith snapped.

'No,' Lumia said. 'No I don't. They don't need mercy. We need mercy here. Your family.'

Abu Laith, who was not a diplomatic man, gave up answering and let her shout as he kept packing food into the bag, the children scampering around offering unhelpful advice.

Beneath the surface, Abu Laith was very afraid. The airstrike had been so close, and the metal cages could have been weaponized into shrapnel, cutting the animals to pieces. With all he had, he hoped that he was wrong, and that the lion he had raised with a bottle was alive.

He stood up. The sky was darkening, and he could hear the army's artillery somewhere in the distance behind him. In front, just a few streets behind the zoo, lay Daesh. They were stuck in no-man's-land between two highly armed enemies. He knew he had to go now. Daesh could push back at any time, or the army could sweep forward, making the street impassable with machine-gun bullets and mortar rounds. For now, he and his family and animals were in the eye of the storm, and the night was strangely quiet. It would not stay that way.

Abu Laith bided his time in the back room until nightfall, when the shelling intensified. The children were swirling about the room, all thoughts of Abu Laith's mission forgotten in favour of their toys. Even Lumia had stopped being angry, as she sometimes did, very suddenly and with no explanation, like a deflating balloon. Abu Laith crept out of the living room, picking up his sack of food as he went.

He walked into the courtyard. The air was dark, and the front of the house was quiet. He listened in vain for Zombie's roar, and opened the gate. The street looked even worse in the moonlight. Five days ago it had been a dusty lane adjoining a zoo; now it was scattered with rubble, great chunks gouged out where the shrapnel had hit walls and pavements. Now the distant artillery made the air roar, punctuated with bursts of small arms fire. It came from the other side of the zoo. But the street itself was calm.

The gates were open, Abu Laith noted crossly, as he half-ran up the street. There was nothing to stop the animals escaping, or to stop anyone coming in to steal them. Those peacocks were worth hundreds of dollars each, probably.

Abu Laith went in through the gates, and stumbled at the edge of a bomb crater. It was as if a great fist had punched down on the zoo. Everywhere there were gaping holes, some deep and others shallow. The Daeshi bodies were gone, and the trees were broken, white splintered trunks gaping in the moonlight.

He heard something moving in the bushes, and stopped. With a rush of fear, instant and biting, he wondered if the animals had all escaped, or if the Daeshis had let them out.

Zombie, he knew with a zookeeper's certainty, would never attack him. He had fed him buffalo milk from a bottle to make him strong, and played with him like he was a puppy. But Father – and especially Mother – were another problem. They were starving lions, raised in captivity but probably, Abu Laith assumed, able to hunt. He knew from watching lions chase zebras on the National Geographic channel that they could run very fast when the urge came.

He decided to take his chances. The animal enclosure lay ahead. Abu Laith ran through the gate, towards the rabbit cages. The doors of the hutches were open, and they were empty. It had been more than two weeks since anyone had fed the animals, and Abu Laith supposed they would have died if no one had stolen them.

He kept on straight towards the ostrich cage. Even from a distance, it was clear there was something wrong. Abu Laith knew that ostriches never slept sitting down. Inside the cage, there was only a dark mass on the floor.

He was filled with a sorrow so deep that he felt helpless in the face of it. It was his fault. He had called in the army. They had known that there were Daeshis in the zoo because of him.

The ostrich feathers were oily under his touch. Both the adult and the children were slumped on the floor of the enclosure, bodies hidden under their plumage. They stank. They must have died from the sound of the bomb, he thought, rather unscientifically.

Abu Laith walked towards Lula and Zombie's cages. He knew they were dead. They hadn't eaten or drunk in weeks. They couldn't have made it, he knew.

Zombie was lying on the ground, head on his front paws, perfectly still. Abu Laith sank down to his knees and reached through the bars, his hands ready for the cardboard cold of dead flesh.

Instead, there was a warm rising and falling of breath. 'Zombie,' Abu Laith shouted, joy rising in him like hot water. 'You're alive.'

To his right, Lula roared, and Abu Laith felt as if he could dance. He ran over to her cage. She didn't look well. Her grey fur dry and brittle, she was twisting herself around in tiny circles, uttering faint cries. The corner of her cage was full of gravel, kicked aside as she threw herself against the bars. Warda cowered behind her, whimpering. Next door in the lions' cage Father lay, as usual, still on the floor, but at least he was breathing. Mother sat impassive, eying her human visitor.

Going from cage to cage, Abu Laith hummed quiet encouragement at the animals as he threw in handfuls of food – chicken and meat for the lions, rice and beans for everyone else. He stooped low in front of Lula's cage, and stuck his hand inside to stroke her fur. She had lived through war. She had lost her mate, and almost lost her child. He wondered if this time, life would have broken her.

'Everything will be OK,' he told her, not quite believing it. 'Don't worry.'

He crouched down, and waited for the bears to calm. Warda was behaving very fearfully, hiding behind his mother's bulk when usually he would have run up to sniff the zookeeper's hands. Abu Laith waited. The mortars had started again, as they did every night, and he could hear the crashing from both armies nearby. After a moment, Warda moved towards him. The three-legged cub had a deep fresh shrapnel wound in his side, blood dripping on to the floor.

Abu Laith tried to beckon Lula over but, still rigid with fear, she pulled her cub back to her chest. Abu Laith stood up, choking back tears. He opened a bag of rice and beans and threw it on to the floor of the cage for them. There was nothing else he could do.

He hurried over to the monkey cage, listening hard for their chattering. Two of them scampered over when he got to the cage, grabbing the mesh with their tiny-fingered hands. They looked skinny, but very alive. Something had blown apart the corner of the cage, and there was a large hole amid the mess of wire.

'Where are your friends?' asked Abu Laith, politely. Nusa,

the oldest female, and her child weren't there, nor was the baboon, Marwan's favourite and the scandal of the women of Mosul. 'Where are they?' Abu Laith asked, walking back and forth in front of the cage, eyes raking the floor. 'Did they run away?'

There was a dark patch in the corner of the cage, down by a pile of branches that served as the monkeys' living space. Abu Laith stared at it, horrified. The baboon, all two feet of him, was lying still on the ground, arms splayed out. He was manifestly dead. The baboon would never again expose himself to visiting women, or swing among the branches chattering to Marwan. Abu Laith, heart broken, supposed he had been killed by shrapnel.

The eldest male monkey hopped on to a branch. In the place where his tail should have been, there was a crusted stump that twitched disconsolately. It was a terrible thing, Abu Laith thought, for a monkey to lose its tail. He knew, from National Geographic, that they used them to balance while climbing trees. 'I'm sorry,' he said to the monkey, who looked extremely cross.

The other monkeys must have escaped through the hole in the mesh, he thought. Looking up, with a rush of relief, he saw Nusa and her child sitting on the roof of the cage, looking down at him with vague interest. For now, there was nothing Abu Laith could do. The mortar rounds were landing not far from the zoo, and he knew he had to leave. Feeling like the worst kind of traitor, Abu Laith took a last look around the monkey cage. He poured some rice into their food tray, and topped up their water from a bucket.

Then he ran back towards the house, empty plastic bags crumpled in his hand.

34

Abu Laith

Luay had been sitting on his mattress in the living room all morning when he heard the beeping noise. It came from the road: loud, insistent, and – to a mechanic's son – immediately identifiable as the sound of a parking guide on a reversing car. He got up and went to the door. No cars had gone down that road since the airstrike a week or so before. The fighting had grown a little quieter as Daesh moved back towards Mosul city centre, but the road was still scattered with rubble and mud that no normal car would ever be able to drive through. All this passed through his head in an instant as he looked out across the courtyard, listening to the beeping noise.

Then came a snatch of music. He hadn't heard music in public since Daesh had come, two and a half years ago. It came only for a second, then disappeared, like it had been playing out of the window of a fast car on a road not too far away.

Crouching down, he slid over towards the gate. He unlatched it and pushed – gently as a breath – with his fingertips. Through the gap, he could just see the rear end of a large armoured vehicle, which was reversing down the street away from him. There was a machine-gun turret on the roof, with a black helmet – which was presumably attached to a man – poking out the top. An Iraqi flag – green, black, red and white – flew from the turret. Over the beeping, he could just hear someone shouting on a walkie talkie. Luay watched the car disappear back down the road. Then a grin broke out on his face, and happiness filled him.

'The army is here,' he shouted, as he threw himself back into the living room, where the rest of the family were sitting, engaged in their usual pursuits of nagging, fighting and plotting.

Abu Laith stood up. 'Where did you see them?' he asked. 'Are you sure it's them?'

As the army had drawn closer, rumours had swept through the neighbourhood that Daesh were planning to dress up as soldiers to lure out the people who had opposed them. When civilians came streaming into the streets to welcome the army, they would launch suicide attacks and kill them.

This was not, Luay was sure, what he had seen. 'Seriously, it was them,' he said, near-jogging with excitement. 'They drove down the street in a Humvee, and I could hear music playing from another car.'

Abu Laith beamed. 'I'm going to the zoo,' he said, and was on his way out the door when Lumia stopped him.

'You're not leaving us,' she barked. 'You're staying here until we know for sure the army are here, and that the Daeshis are not coming back.'

Abu Laith was mutinous, then resigned, as Lumia glared him down. 'Fine,' he said. 'But the minute we're liberated, I'm going to see the zoo.'

'Yes, yes,' said Lumia, who couldn't be bothered to argue any more, nor entertain the reality that Daesh might actually be gone, which just seemed ludicrous. 'We need you to protect us.'

As they argued, Luay was already running out of the house and into the garden. He knew it hadn't been a trick. It had been too real – the flag, the walkie talkies, the music. The impossible idea that they might really be free, that the war was over and Daesh had gone, began to take root in his mind. Feeling unusually bold, he walked over to the gate and swung it open. The road was empty. For the first time in weeks, he took a step into the road. The gate swung shut behind him, and he crept along the wall towards the main road, moving low and fast in case there were snipers around. Those were risks he had never known before the war, and that he now thought of automatically, as if he always had. There were no shots. He was excited and terrified in equal measure.

At the end of the street, where the main road began, he saw people walking, not running. A young guy he knew vaguely from the neighbourhood came up to him. He was grinning, and Luay, when he saw what he was holding, grinned too. 'You're fucking kidding me,' he laughed at his friend, who was waving a packet of cigarettes in his face. 'You've got to be fucking kidding me.'

'No way,' the other laughed, brandishing the cigarettes. 'It's happened. These are mine.'

Luay laughed with him, and at the crowd of men who had gathered around him with boxes of cigarettes, trying to flog them to him. He hadn't seen cigarettes for more than a year, not since Daesh had clamped down on the black-market smugglers, beheading them when they could.

'Check it out,' his friend shouted over the babble. 'I've got tobacco for the hookah pipe too. We'll smoke tonight.'

Luay laughed like he hadn't for years, since he was last a real teenager, rather than the terrified man that Daesh had made of him. As he walked down the street, he stared around him as if he had come from outer space. There were soldiers parked all the way down the street, rousing army songs playing out of their trucks and armoured cars, flying the Iraqi flag. They didn't look threatening, like they had before Daesh came. They were laughing and waving at the pedestrians. People were walking up the road on their way out of the city, some holding bags, others pushing carts. They looked tired and drawn; some were injured and being carried by their relatives, or pushed along in wheelbarrows. Only the children seemed happy, though the whizz and thud of the fighting was still close. Luay tried to pull himself together and look at his feet – he had to make sure he didn't tread on any mines. Daesh would have buried them here, he knew, to stop the advance. It had been all over the news.

But as he walked, he still looked up. There were women on the street, either strolling around like him, or heading out of the city with their families. One young woman in front of him pulled up her *khemmar*, letting it fall over the back of her head. Her face, white as milk, shone in the winter sun, her eyes closed, drinking it in. Luay stopped on the street and gaped at her. He hadn't seen a woman's face, other than his sisters' and his stepmother's, for years. All around him, women were pulling up their *khemmars*, turning their sun-starved skin to the sky. They all looked beautiful, he thought, as he walked down the road, feeling truly free, though the guns were still firing not too far away.

'It's happened,' shouted Luay half an hour later, as he barrelled through the gates of the house. 'Look, come and see.' The children, hotly pursued by Abu Laith and Lumia, rushed into the kitchen, where Luay had thrown a brace of plastic bags down on the counter.

'Vegetables,' said Lumia, as she picked an onion out of the bag. 'Where did you find them?'

'We're free,' Luay said. 'There were soldiers in the street, and women with their faces showing, and everyone was smoking.' He brought out a carton of cigarettes, opened a box and lit one. 'See,' he laughed, as Lumia cackled. 'See?'

'Give me one of those,' shouted Abu Laith, who was almost jigging on the spot. He took a cigarette and smoked it in a few greedy gulps. 'I'm going to the zoo,' he said, and this time no one tried to stop him. He had spent many nights now sleepless, ears pricked for the sound of Zombie's roar, but none had come, and he was terrified and ready for the worst. But he needed to know.

Dizzy with the unfamiliar hit of nicotine, he bolted across the garden and opened the courtyard door. He stopped dead.

There was a man, not a metre away, wearing a tight black uniform and a helmet. He was carrying a gun, and shouting at a group of other men in an armoured car, which had an Iraqi flag flying from the top. There were another two vehicles behind it. The soldier jumped at the sound of the gate opening and turned to Abu Laith, who skittered backwards.

'Who are you?' barked the soldier. 'Are you Daesh? Where are the Daesh fighters?'

Abu Laith steadied himself. 'We're not Daesh,' he said. 'My name is Abu Laith. I'm the zookeeper.'

The two men stared at each other for a moment. The

soldier had an Iraqi flag patch sewn onto the shoulder of his uniform. His comrades were filing out of the Humvee. None of them had their guns drawn. 'We never supported Daesh,' added Abu Laith, for good measure. 'My daughter works for the military intelligence in Baghdad. We support the Iraqi army.'

The soldier at the front, the one who had jumped, seemed satisfied by this. 'OK,' he said, in a Baghdadi drawl. 'Can we come in? We're just checking the houses.'

Abu Laith opened the door with a sinking feeling. He knew that Lumia would be scared of the army. Enough of their relatives had disappeared at checkpoints for her to fear them almost more than Daesh. The extremists were a known quantity. If she stayed quiet and covered up they probably wouldn't bother her. But the army, and the militias who worked with them, could rape, kill, steal because they felt like it.

As the soldiers walked into the house, he dashed ahead to tell her. 'Lumia,' he hissed. 'The army are here. Get the kids into the back room.'

His wife looked terrified. 'Are they the army or the Peshmerga?' she asked, ushering the children away.

'Army,' said Abu Laith, perplexed by why his wife needed to know whether they were being liberated by Kurdish or Iraqi government forces. 'What the hell does it matter?'

Lumia was breathing short, calming breaths. 'I've got a plan,' she said, quickly. 'If they were Peshmerga, I was going to tell them that my grandmother is Kurdish, but since they're the army I'm going to say I'm from Baghdad.'

It took Abu Laith a moment to process this before he burst into laughter, as his wife kept chattering. 'I'll tell them my aunt is married to a Shia,' she said. 'And then they won't behead us, because they won't think we're Daesh.'

'You're a devil,' said Abu Laith, wiping his eyes. 'They're not going to behead us. Come on, let's go and see them.'

Together, they walked into the yard where the soldiers were waiting. Abu Laith was about to invite them in when Lumia sprang into action.

'Welcome,' she shouted, in an abomination of a Baghdad accent she had to have learned from a TV show. 'Come in, please, I am your sister.'

The soldiers, most of them dark-skinned southerners, looked at each other. Abu Laith was trying so hard not to laugh he thought he might bite his lip off.

'Where are you from in Baghdad?' one of them asked. 'What are you doing here?'

Lumia was momentarily flustered. 'Where are *you* from in Baghdad?' she asked the soldier, softening all the harsh Moslawi Qs into Gs.

'Al-Adhamiyah,' the soldier said.

'My grandmother lives there,' said Lumia, who had barely heard of the place. 'What a coincidence. Welcome, all of you.'

The commander, who had seemed decidedly confused at first, pulled himself together. 'All of your ID cards please,' he said, as some of the men filed into the house. 'We've been here since dawn, but we didn't want to come in and wake you up too early.'

'That's so kind,' said Lumia, as Abu Laith went for the ID cards. 'But please, we have nothing, please don't destroy the house. We're not Daesh, I'm from Baghdad.'

'Don't worry,' the commander said, Lumia remembered. 'We just need to check that there are no weapons here.'

Abu Laith came out with the ID cards. 'Officer,' he said. 'You might find some electronics on the top floor, but they're not bombs. I'm a mechanic. They're just tools to fix cars, and spare parts.'

At this, the commander's face set. 'Where are they?' he said, sharply. 'What kind of tools?'

There was a crash from inside the house, and Nour screamed. Lumia ran in to the living room, Abu Laith following her. One of the soldiers had turned over the sofa, looking for weapons behind it. Nour was sitting on the other sofa, baby Shuja swaddled in her arms. Both of them were crying.

'What are you doing?' shouted Lumia. 'We don't have guns. You don't need to do this.'

The soldier grinned at her. 'Are you sure?' he said, and the others laughed.

'Look as much as you want, brothers,' said Abu Laith genially, and grabbed Lumia. 'Stop that,' he hissed in her ear. 'You'll make it worse. They'll think you're hiding something.'

She nodded, and Abu Laith saw she was very afraid.

'I'll make tea,' she said, and walked off to the kitchen.

Ten minutes later, the house had been searched in a rather perfunctory way and half a dozen soldiers, a zookeeper and a fake Baghdadi were sitting in the garden drinking tea.

The atmosphere had lightened considerably since Abu Laith had borrowed one of the soldier's phones and called Dalal, who had managed not to cry with relief that her family was alive, and had lectured the soldiers on the importance of treating them well.

Lumia was a little calmer, though still wary, and still extremely Baghdadi. 'God bless you, God protect you,' she repeated as she gave them all their tea in the garden. She was still terrified that they would take her house from her. They were all-powerful, armed with American weapons. If they told the family to leave, they would have to go. She kept the children shut in the back room, Nour and Shuja with them, and hoped the men wouldn't turn nasty.

Abu Laith, who wasn't scared of soldiers, having been one himself, told them about Abu Hareth threatening to slaughter him like a sheep. 'But he's a pimp,' he shouted, as the soldiers laughed. 'He's a hypocrite, like all those mullahs.'

'Listen,' said the commander, after the laughter had died down. 'We need your help. What did Daesh leave behind in the zoo?'

'There was a lot of ammunition and a pile of mortars,' Abu Laith said. 'I can show you where it is.'

Lumia wasn't happy. There was, as she pointed out to Abu Laith, still fighting around the zoo. This was the army's front line, and Daesh would be trying their hardest to drop mortars on their heads, despite the no-man's-land between them.

'Don't worry,' said Abu Laith, importantly. 'The army needs me.'

With an air of decisiveness, he marched through the front gate. The soldiers followed him into the zoo.

'Are you sure there aren't IEDs here?' asked one of the soldiers. They were all walking suspiciously far behind him, treading in his steps.

Abu Laith turned around. 'I think I'd know if there were mines in my zoo,' he said, a little sniffily, despite knowing nothing of the sort. 'It's clear, don't worry.'

He showed the soldiers the place where the mortars had been, and the gap under the carousel where Isis had hidden their ammunition. But he could barely concentrate on the task at hand. Fear for the animals was gnawing at him like a rat. He needed to see Zombie.

Leaving the soldiers behind, he made for the animal enclosure, deeply nervous. He didn't know what he would find there. It had been a long time since he'd left them the food after the airstrike. He didn't know if it was even physically possible that they'd survived.

Rounding the corner, his heart flipped. Zombie was standing up in his cage, looking straight at him, rangy but alive. Abu Laith whooped and ran up to him. Next door, Mother was walking in tight circles in her cage. Father was asleep, but breathing, he saw. Next door, Lula was standing in her cage.

'Zombie,' Abu Laith shouted. His eyes blinked with tears of relief. 'You're alive.'

But before he could go much further, something caught his eye. Lula was digging at the ground in a very strange way. He went over, and was almost sick. Lula was digging a hole, and next to her was a desiccated pile of something that had – Abu Laith knew immediately – once been a bear cub.

'Warda,' he breathed. 'Lula.'

Abu Laith wept for the bear cub, with its soft belly and three legs. Lula was still digging frantically. She wasn't making any progress through the hard, stony floor. Abu Laith understood, instantly, that she wanted to hide his remains.

Behind him, the soldiers walked into the zoo. The commander came over to Abu Laith. He was staring from Lula to Zombie, to Mother and Father.

'Aren't you people hungry?' he asked, incredulously. 'Why didn't you kill them for meat?'

It was too much. Months of near-starvation, a bear cub

dead, and now this insult. Abu Laith, with tears still in his eyes, burst into a blank fury. 'Why didn't we eat them?' he yelled. 'You don't eat animals who have earned your respect. We all went hungry to keep them alive. That's what respect is.'

With a glare, Abu Laith calmed down, and remembered his duties as a host. They would probably like to see the lions first, he thought and brought them over to their enclosure. Zombie, who had been yowling, stopped when he saw Abu Laith standing in front of his cage. But Mother was still screaming like a demon, twisting around in her cage and raking her claws across the bottom of the cage. She had lost her mind, Abu Laith thought. Father was asleep. It was a bit embarrassing for the houseproud zookeeper. They moved on quickly.

'These are the monkeys,' he said, standing in front of their cage. When he took a closer look, however, he saw that it was empty but for the remains of the baboon in the corner. The surviving monkeys had escaped. Tears sprung to his eyes again. The baboon was dead, but he would not let the others down. He would find them, wherever they were, and bring them home.

'You need to be careful,' said one of the soldiers. 'Daesh have left sleeper cells behind. They might attack you because they know you reported to us.'

'I'm not afraid,' Abu Laith later remembered telling them. 'I'll beat up anyone who tries to bring Daesh here.'

The soldiers, who had heard the same thing from almost every civilian they had met in the battle to retake the city, were probably not convinced. No civilian in Mosul was above suspicion. It was odd, some of them joked, how every single local had lived under Daesh so long despite hating them. But this old man might be one of the good guys. For one thing, he kept asking them where he could buy whisky. For the other, his daughter was in military intelligence, and by all accounts, she was blonde and terrifying.

35

Abu Laith

The emergency committee had assembled by 4 p.m. Abu Laith stood in the sun-speckled courtyard in front of the children, a manic glint in his eye. All of them were armed to the teeth with sticks and fruit peel that they might use to entice or encourage escaped monkeys to move.

'I don't care what you've heard,' said Abu Laith. 'But there is only one proper way to catch a monkey.' The courtyard was rapt and still. 'They might run around all night,' Abu Laith continued. 'But in the morning they get tired. So what we're going to do is find them, and wait. Then when the morning comes, they won't be able to stay awake. That's when we go up to them, slowly as we can, and we'll catch them.'

The rescue team readied itself for the expedition, with rather more shouting and shoving than necessary. Geggo, alone among them, carried only a bag of crisps. He had bought them that morning from the corner shop by the zoo, with money saved – or rather, found around the house and appropriated – over several weeks. They would, he thought, be a good reward for any monkey they found. Monkeys liked crisps even more than he did.

With Abu Laith at their head, the children marched through the door and turned right through the peeling backstreets of their neighbourhood. As they walked, curtains twitched alongside them. Many of the neighbours, Abu Laith knew from bitter experience, were light sleepers easily woken up by lions roaring, dogs barking or zookeepers singing after a few whiskys. They did not like him one bit.

'Idiots,' he thought, brimming with contempt for the watchers behind the curtains.

Even so, he hoped that the monkeys hadn't got into their houses. They would probably take a dim view of that. As he walked past the zoo he kept an eye out, as he always did, for a young woman by herself. He had promised Marwan that he would try to check if Heba, his fiancee, had come looking for him – and try to pass on a message that he hadn't abandoned her. But it was, of course, pointless. He had no idea what she looked like, and she had never met him before. Still, he looked.

Abdulrahman was on high alert. His eyes critically raked every fence and bush that might hide a monkey. 'I guess they'd go somewhere with food, wouldn't they,' he pondered, scouting the tree-line.

'Or water,' said Luay, who was carrying a large net.

As they searched, the children scattered across the neighbourhood, moving with deadly seriousness. The clock, they knew, was ticking. Monkeys couldn't survive for long without food.

Help soon came in the form of their neighbour Abu Rama, a mechanic of easy temperament who had a garden full of orange trees, which bloomed heavily in winters like this.

'Assalamu aleikum,' said Abu Laith cheerily, when they saw him walking up the road towards them.

'Wa aleikum assalam,' said Abu Rama. 'I think you'd better come with me. I went into the garden last night, and those monkeys were eating everything we had.'

They walked for a while towards Abu Rama's house, Abdulrahman steeling himself for the encounter.

'They've eaten the dates that we have saved up, too,' Abu Rama said, conversationally. 'As well as almost all the oranges.'

Abu Laith was unapologetic. 'Don't worry,' he said. 'We'll have them back soon.'

It was abundantly clear which tree contained the monkeys. The remnants of a dozen oranges lay below it, smashed into pulp. As Abu Laith approached, he could see their long tails sinewy among the branches.

In the tops of the trees, ignoring them all, were the monkeys, munching on a feast of Roman proportions. Their hands were full of orange flesh, and their mouths running with juice. The three of them sat with their arms resting on their distended bellies, pleasantly replete. It was clear they had no plans to leave.

Abu Laith walked slowly over to the tree, peeling a banana. Stretching out his hand, he called out to the monkey closest to him. Usually, she would have run over to see what he had for her. Now she was stuffed and lazy.

Munching slowly, she eyed him for a second, then rested back into the crook of her branch.

It was as he had expected. He turned to the rest of the group. It was getting dark and soon, he knew, the monkeys would be livelier than ever – zooming around the trees high on fructose. They would never be able to catch them in the dark.

'We'll have to wait until morning,' he said. 'Then we'll be

ready for them. At least we know what to bring next time: oranges and dates. That's what they like.'

The chastened group turned towards home. Abu Laith, alone among them, was brimming with enthusiasm for the task ahead. He knew from National Geographic that monkeys liked coconuts even more than they liked oranges, bananas and dates. Slowly, a plan was beginning to form.

At home, Abu Laith set to work. First, he took out his old toolbox. Inside lay a coil of thin cord that he had used in his mechanic's shop. Next, he went to the bedroom to find Lumia's coconut perfume, expensively procured before the arrival of Isis.

Back in the living room, Abu Laith set to work spraying the perfume over the cord, which he cut into suitable lengths, ready to be used by his volunteer trappers. Then he tied loops in them, so that they might be slipped around the monkey's waists and tightened. When he had finished, he put them away and went to bed, satisfied with his genius.

The next morning, Abu Laith was up and moving before the sun was fully in the sky – shoving past the children to the door with his newly oiled cords. Behind him trailed Luay, who was carrying a net, and Abdulrahman, who had barely slept he was so excited.

Silence was absolute, enforced by a general understanding of the seriousness of the situation.

Abu Rama stopped them on the street outside his house. 'They've gone next door,' he said.

The house in question had been abandoned since Isis came, but the family knew it well. Abu Laith's oldest son had lived there for a while – they'd had his wedding party on the second floor. The door was open, and they climbed up the stairs. Moving quietly, they fanned out across the house. Luay and Abu Laith went into the master bedroom, which had been decked out like a honeymoon suite. There was a big mirror and photos of the bride and groom on the sideboard.

'Wait,' whispered Luay. 'Look.'

He pointed at the curtains. There was an unidentifiable bulge around the bottom by the floor, and they were moving in the windless room.

'There,' breathed Luay.

Abu Laith stopped the others and walked forward slowly, like a hunting bear. As he went, he pulled the chords out of his pocket and laid them one by one on the floor, each tied into a loop that could be tightened at will.

'I'll lure them in, then I'll tighten the cords around their waists,' Abu Laith told his sons. 'Not around their necks. That would be dangerous.'

Abdulrahman looked at the monkeys who were hidden in the curtains, visible only if you looked at them from the right angle. One was big; he was the father. His stump of a tail twitched as he sat, a half-healed shrapnel wound still visible on his leg. One was the mother, holding her baby. It would probably be very hard to get them into the loops, he thought.

For five minutes, Abu Laith's sons waited as their father tried to tempt the monkeys into the coconut-scented rope coils. Soon, it was clear to everyone except Abu Laith that not all monkeys have an equal love for coconut oil.

'Shall we try the net?' asked Luay, advancing on the curtains, holding it out in front of him.

Abdulrahman, gleeful, followed. Abu Laith joined them. Together, they pulled out the net so that the curtains were shut off on three sides, and blocked by the closed windows on the other.

Slowly they stretched forward, each moving within an arm's reach of the dozy creatures. Then, in a lighting strike, Abu Laith threw himself forward. A shriek, and the curtains exploded into pandemonium.

Abdulrahman had grabbed his monkey with the first try, but fumbled in his elation, letting a stringy arm spring out and make a bid for freedom. Luay's target had woken up before he could reach it, and scurried up the curtains. The young man tried to follow, grabbing for the monkey.

Within a few moments, Abu Laith had the mother and baby in his arms, trying to pull a cord around the older one's twisting waist. He gave up, and wrapped them gently in the net. Together they observed the debacle in front of them.

Luay was launching himself towards the top of the curtains, grabbing for the father monkey who was strutting around as his family lay swaddled in the net, wrapped like sausages. With a leap, he tried to grab him, but missed. Another try, and he had him.

By the time the sun was high, they were traipsing through the zoo carrying their bundles on their way to the monkey cage, which Abu Laith had repaired with a piece of old chicken wire. Luay put the mother and child inside and the father, who had followed them, jumped in after. Soon, the monkeys – who seemed no worse for their adventure – were scampering around the cage again, adjusting to their battered, but familiar, surroundings.

Abu Laith, beaming, stood outside the cage, shouting at anyone who would listen. 'There's only one way to catch a monkey. Only one. Wait until the morning and grab them with a net.'

As the others traipsed home for breakfast, only Geggo was left. He sat looking at the monkeys in the cage as they chewed their fruit peel. The mother, his favourite, was looking at him. 'She wants a crisp,' he thought, and opened the pack. He pulled one out, and held it through the bars. She took it.

The other monkeys saw what was happening. One by one, they ran over to ask Geggo for crisps. With a great sense of purpose, he handed them one each. Only a few were left when a minor scuffle broke out. In surprise, Geggo's hand closed over the crisp he was holding.

Then something warm clamped on to his hand, and pressure closed over the top of his thumb. The pain came seconds later, as blood started to pour. He screamed and the monkey let go. The chicken wire caught his hand as he tried to pull it back.

There was blood pooling next to his thumb. He felt helpless and betrayed. Terrified, he stood and screamed, not knowing what to do.

Back at the house, Luay was indulging in his usual pre-Daesh pastime – lounging outside the courtyard gate smoking narghile – when he heard a yell from inside the zoo. He knew Geggo was playing there. He bolted through the gates to the animal enclosure, fearing the worst.

His little brother was running about outside the monkey cage, a crying whirl of bloody clothes. 'What happened?' asked Luay, scooping him up from the floor.

Geggo could barely speak, sobbing into his brother's chest.

'Who was it?' Luay asked, scrabbling around his brother's clothes in a panic, looking for a wound. 'Who hurt you?'

Geggo wailed, and held out his hand. Luay grabbed it, the blood seeping through his fingers as he turned it over. 'Is it only this?' he asked, as the fear subsided.

'It was the monkeys,' his brother sobbed. 'I thought they were my friends.'

Abu Laith came out, summoned by the commotion, and caught the end of Geggo's explanation. He took his hand, looked at it briefly, and dropped it. He roared with laughter. 'That's what I'm telling you,' he said to Geggo. 'This is the law of the jungle. That's what you get when you trust monkeys without earning their respect.'

And Geggo, horrified, wailed again.

36

Hakam

HAKAM TOOK CAREFUL AIM THROUGH HIS SNIPER SCOPE and fired on the advancing soldiers, killing them all in one burst of fire.

'Take that,' he laughed, slamming his laptop shut as the others in the living room moaned.

He was playing *Call of Duty* with his cousins on a shared network he had rigged up. Behind them, his cousin's grey parrot – Cusco – occasionally wailed like a baby, bringing the women of the house crashing into the room, before scolding him for scaring them again.

Three weeks had passed since the Isis fighters had driven them from their hideaway in the bathroom. The family had fled to Hakam's grandmother's house on the outskirts of the city, where they had joined another dozen of their relatives. There were so many of them that Said and his family were given a guest cottage in the grounds.

The house was stocked with food and had high walls that protected them from the gunfights that occasionally broke out on the streets. It was far enough from the fighting that it didn't really feel like they were in a war at all, despite only being a few miles from the front lines of the toughest urban battle since the Second World War. They could walk in the garden in the sun. Using ropes slung over the ruined northernmost bridge – which lay nearby – for support, they could walk over to the western side to replenish their supplies, lugging back oil drums and water.

They were relatively safe, but the food was plain, and the boredom grew thick, and everyone wanted to go home.

Hakam and his younger cousins, Mohammed and Abdullah, tried getting into an exercise routine. All of them were physically almost unrecognizable; Hakam's arms were nearly as thin as Hasna's. The pointless waiting made everyone tired. After a few attempts at exercise, Hakam and and his cousins gave up, disappointed at how weak they had become.

In the evenings, Hakam sat with Omar, his favourite uncle – an ebullient man in his mid-thirties with two small children and a rolling laugh. Before the war, he had run a food store, and had a minivan that he rented out for deliveries. Since Daesh had arrived he had – he always joked – spent most of his time looking for cigarette smugglers to fund his twenty-a-day habit.

Hasna tried to keep her spirits up for the kids – most of them her cousins – who scampered about the house. Whenever she could, she would walk from the cottage to the big house to play with them. But most of the time, they slept and waited. It was, Hasna thought, like being dead while you were still alive. Every day, the fighting grew closer.

No one was sure whether the army would come this way. They were so far down the eastern bank of the Tigris that it barely counted as Mosul, and they all hoped that the district would not become a war zone, and that any militants nearby would just make their escape.

In early January 2017, after a few weeks that seemed to stretch into months, a few of the older men, including Omar, decided that the time had come to see what was happening around them. Emerging from the house in a group, they scouted the neighbourhood, leaving Hakam and his cousins at home where they couldn't be mistaken for Isis fighters by the army or for soldiers by the militants. There wasn't much going on, they reported, but they continued cautiously to explore the area.

One morning, Omar had run out with another uncle and a neighbour, after they heard there was a house fire nearby. They wanted to help.

The women made breakfast, wading among the kids and laughing over the screams for attention. Hakam and his cousins were in the living room, watching TV. They had just sat down to eat when one of Hakam and Hasna's uncles ran into the room.

'Omar is dead,' he said. 'He was shot. The others took him in a boat to the hospital in the west.'

Hakam couldn't speak.

'There was a fire down the road,' his uncle choked. 'And we ran down to see if we could help. But then someone started firing at us, and they shot Omar right through his chest.'

For days they mourned, immobile. Omar's wife was deep in shock, but refused to let her grief show to her children. Hakam couldn't think straight, let alone accept his uncle's death.

About five days later, the family came to a decision. Another uncle had gone back into the city to check on his house, which was around the corner from the Zararis' place, and reported that both were still standing and had been recently liberated. He begged them to come home before their house, with its plants and its birds of paradise, was looted.

'There's a few snipers on the streets around here,' said one of Hakam's cousins, as the family met to discuss their plans. 'But if you're careful, you can get to the road.'

'Too many mortars,' said another relative. 'There's no way you'll make it.

'The problem is the road, it is blocked, the Americans destroyed it so that Daesh can't flee. You can't drive through.'

Then Arwa spoke up. 'We'll walk it,' she said. 'We need to go back home.'

The next morning, the Zarari family each packed a backpack and walked into the courtyard. A dead light hung in the air – the sun veiled by dust. From a few streets away came the sounds of fighting.

Said went first, sticking his head out of the front door and checking if the coast was clear. He would go at the head of the group, followed by Hasna, then Arwa and Hakam. They would walk in single file, spaced far apart, walking fast and crouched low. That way, if one of them stepped on an explosive device, it wouldn't blow the others up, and they would offer a less tempting target for snipers.

Said was waiting on the other side of the gate, hustling his family quietly into the abandoned street. They turned down the road towards the city and half-walked, half-ran through the ghostly stillness.

The houses around them bore bullet and shell scars from the battle. They seemed empty behind their gates, but every flicker of sun on a window, or waving tree branch, made them move faster.

'Look down,' said Said. 'Remember to watch your feet.'

As Daesh had retreated, they had hidden thousands of IEDs throughout the city – under roads and behind doors and even in childrens' toys. When they could, they booby-trapped all the streets they retreated from, to stop the army from following them.

Hasna was so intent on looking down that she didn't see the head until she was almost on it. It lay in the road, the body nowhere to be seen, tossed among some rubble. It didn't have a face. Maybe the dogs ate it, Hasna thought, as she hurried on, trying not to retch at the smell. There had to be more bodies here: the air was thick with the smell of corpses and the flies buzzed black, a great blanket in the air. Hasna wrapped the end of her hijab over her nose and mouth.

Crouching, scanning the buildings around them, Hakam, Hasna, Said and Arwa ran forward, clutching their bags close. They were in no-man's-land, potential targets for both the army and Daesh.

What had once been a road was now a dirt track piled with rubble, glass and the burned wrecks of cars. It looked, Hakam couldn't help thinking, like the war zones in Counter Strike, which he used to play with his cousin after school when they were kids.

Said led the way, stopping every time they reached the corner of a building to look around it, checking for Daeshis. All of them were pretending to be calm. They had survived the fighting, but there were still drones in the sky and IEDs in the ground, and death could come from anywhere.

An hour later, they reached an intersection, and the first living people they had seen on their journey. All of them were escaping from somewhere – bags tumbling from their arms, children clutching trouser legs and wailing. The Zararis walked among them, dirty and thin as all the other Moslawis walking through the intersection that day, so tired all they could do was stare at their feet and hope they wouldn't die.

Said ran forward to look around another corner, and stopped. 'It's the army,' he said. 'They're here.'

Hakam couldn't even bring himself to care. He was so used up with grief and by the constant danger, that he just kept walking. They trudged around the corner and saw a Humvee in front of them with an Iraqi flag on the top. Soldiers were standing next to it, guns at the ready and armoured in their helmets and tac vests. They were jumpy, terrified of suicide bombers. But they weren't searching anyone, and the downtrodden civilians walked past them into liberated territory, barely giving them a glance.

'Take that off your face,' shouted one of the soldiers as the family came close, and Hasna realized he was talking to her. She was so used to the *khemmar* that she had forgotten that she'd wound her scarf around her face. They probably thought she was a Daeshi.

'Sorry,' she shouted, pulling the scarf free. The soldier let them pass through the lines.

'Walk on the street,' one man with a shaking madness about him screamed at his wife. 'Not on the pavement.' She obeyed, but he kept shrieking at her, eyes wild.

As they walked, Hasna felt her anxiety build. The most dangerous part of the journey still lay ahead. As the army had advanced on the east, Isis had reinforced their positions along the western bank of the river. From there, they launched their home-weaponized drones and mortar fire at anything that sniffed of opposition. The Zarari family, with their backpacks, more than qualified.

To get home, they needed to walk over an intersection at the mouth of one of the bridges across the river, which had been destroyed in an airstrike. They had to cross 50 feet of barren ground, exposed to the gunsights of the Daeshis.

The family half-jogged forward as gunfire cracked nearby. Hakam was trying to strike a balance between watching the ground ahead of him and looking around for snipers. At any moment, he expected the earth to lift under him.

Ahead, Said stopped at a corner. 'We're by the river,' he said.

Arwa moaned and sat down. 'Can we wait?' she asked. Her face was streaked with sweat.

'No,' said Said, gently. 'The drones will get us. We have to go.'

'We'll be almost home once we're over it,' said Hasna. Arwa straightened up and readied herself. They rounded the corner at a run. They'd be protected by the surrounding buildings for another few hundred feet or so, until they made it to the lip of the intersection. They could already see a couple of dozen people gathered near the ruined bridge, waiting to cross.

'There's a sniper on the other side,' one of them called. 'We'll wait until there's more of us. Then we'll run.'

Hakam knew it made sense. They would be harder to hit if they were in a large group. But no one would want to be at the edge of that group, and the Daeshis could just lay a burst of automatic fire that would cut them all down.

Said told them what to do. He and Hakam would go first, drawing the fire. Hasna and Arwa would cross together behind them. More people joined them, more hot, unwashed Moslawis with their possessions in their arms.

'Go,' someone shouted, and the Zarari family and the dozens of people around them all ran for a spot on the other side of the intersection that seemed unreasonably far away.

Hakam didn't have the time or the energy to process what was happening. It was hot, and he wasn't listening for shots, but just running as fast as he could, fixing his eyes on the shell-pocked office block on the other side.

And then he was there, and there was only silence, and the sniper hadn't fired. He hid behind the corner of the building. He could feel Said hanging on to his arm, stopping him from running back out. Arwa and Hasna were 30 feet away, then 20. A gunshot cracked, but they kept running until they collapsed by the side of the office-block wall.

'All here then?' asked Said. 'That was a bit close.'

They flagged down a taxi, which was cruising in defiance of the war around it, and the driver agreed to take them home. As they drove, the family passed houses hit by airstrikes – gutted from above, the windows broken, rebar jutting into the air. By the time they reached the edges of their neighbourhood, almost every house they saw was damaged.

There came a point where they could drive no more. The car lurched around potholes 10 feet deep, and the broken glass kicked up at the windows.

'Stop,' said Said. 'We'll walk from here.'

The family jumped out, and the taxi drove away. They walked the last few streets to their house, not knowing what they would find when they got there. There were no Isis fighters, but no army either. They must, Hakam thought, be somewhere in the no-man's-land between the shifting lines.

The peach courtyard walls were still standing. Said pushed the outer door open, and smiled. The Isis car was gone from the yard.

'It's OK,' he said, as the others filed in. 'We'll be OK.'

37

Abu Laith

It wasn't until about the tenth morning after the army arrived that Abu Laith woke up with the blinding realization that Daesh had actually gone. Worries about the animals, Nour's leg, starvation, the infant Shuja, Warda's death, mortars and mines that had gripped him for weeks had dissipated for the first time. He felt buoyant. Outside the window, the winter sun was shining. The mortars crashed in the distance – far west of them now. Later, he would remember how he had felt reckless with happiness. He would go outside.

'Children,' Abu Laith shouted, bounding out of bed and running through to the courtyard. 'Listen to me. Daesh have gone.'

The children were playing under the grapevine trellis. Abdulrahman detached himself and ran over to his father.

'They've gone,' shouted Abu Laith again, grinning.

'We know,' Abdulrahman said, perplexed. 'They left ages ago.'

Abu Laith wasn't listening. 'Let's go,' he shouted at the children, and clanged the gate open.

They surged after him into the road which was littered with rubble, white in the sunlight. Further up towards the old Daesh checkpoint, the street was full of people and honking cars. All around east Mosul, people were realizing that the occupiers had really fled. People in hiding were coming out, bleary-eyed, into the daylight, ignoring all the warnings about Daesh sleeper cells, emerging from the basements of half-bombed-out houses. Abu Laith was back in his element and looking for trouble.

The plan had come to him in an instant upon waking. Abu Hareth's house was up at the end of the road, near where the Daesh checkpoint had been. The Daeshi had fled a couple of days before the army arrived – packing his children and belongings into a white car and driving west. He'd stopped outside Abu Laith's house and shouted for Mohammed, who worked in the shop next door. Abdulrahman had stuck his head out of the gate to listen.

'Watch my house when I'm gone,' he'd said to Mohammed. 'I'll be back soon.' He'd gestured at Abu Laith's house. 'And don't let that guy into the mosque.'

Abu Laith, when he heard, had called in Mohammed for a debriefing and rocked with laughter. 'That bastard,' he'd cried. 'He's never coming back.'

Now the Daeshi's house had lain empty for over a week, and the army controlled the checkpoint by the end of the road. It was time, Abu Laith thought, to show everyone how the hypocritical mullahs really lived.

A group of men stood outside Abu Hareth's house at the top of the lane. It used to belong to a Kurdish family, who had fled when the jihadis came. For the last two years it had been home to the Daeshi and his two brothers, Moslawi men who had joined Isis. Abu Laith hadn't heard from the Kurdish family since they'd gone, but in the summers – years ago – he had shared watermelons with them during baking hot months, when the tarmac bled and the grass crisped. He'd liked them.

'Assalamu aleikum,' shouted Abu Laith, rounding on the group. The men were all coated in a sheen of dust and looked very nervous.

'Wa aleikum assalam,' they muttered back.

Abu Laith sauntered up to Abu Hareth's gate and made to open it.

'You'd better not be going in there to steal,' said one of the men. 'That's our friend's house.'

'You're friends of Abu Tarek,' Abu Laith exclaimed, shaking all the men's hands. 'The Kurds. I knew them well. How are they? What's their news?'

The men had been sent by the family to check on their property. Proper introductions over, Abu Laith returned to the matter at hand. 'Did you ever meet Abu Hareth? He was a Daeshi and a total bastard. He told the people at the mosque that he would slaughter me like a sheep during Eid. He was living in this house, but he's run away now.'

The men commiserated. 'It looks empty,' one of them said. 'But they could have mined the house.'

Everyone had heard stories of people coming back to their homes, only to open their front doors and die.

'We'll see about that,' said Abu Laith, and threw open the unlocked courtyard door. 'He said he would kill me because I liked drinking whisky and didn't pray at the mosque that I built for everyone here with my own money.'

'Be careful,' someone called.

But Abu Laith was already through the door and in the

house, which felt abandoned. It looked much as it had when the the Kurdish family lived there, except far messier. He could hear the others running into the house after him now, fears banished.

He went into the living room and burst out laughing. 'Come in here,' he shouted. 'You've got to see this.'

On a shelf by the TV was a pack of cards. A narghile pipe, with its glass bowl, stood on the floor beside it. If anyone else had been found with either of these, Daesh would have whipped them at the very least.

Luay ran in behind him and picked up the cards, laughing. 'These are mine,' he said. 'I'm keeping them.'

With the customary mad glint in his eye, Abu Laith ran through the house, opening the doors as he went. He saw a transformer that he had loaned to the mosque, and a carpet that he had bought to keep the worshippers warm in the winter. It had cost him 10,000 dinars, and this pimp had stolen it. 'See,' he declared. 'The mullahs even stole things from the mosque. And these are the people you were all praying behind.'

At the end of the corridor he came to a bedroom with a large mirror in the corner. Abu Laith started tearing open the drawers, one by one, and the wardrobe. Finally, he found what he was looking for. 'The lousy infidel,' he shouted, as he held up a lacy red bra that clearly did not conform either to Abu Hareth's chest size, nor the Islamic State's appointed dress code.

One of the crowd burst into the room, and Abu Laith threw the bra at him. It landed on his face, and he nearly fell over laughing. Abu Laith had found another drawer full of women's underwear, some of them no more than wisps of chiffon covered in coins – meant for belly dancing – and was ripping them out by the handful, throwing them into the air. 'God protect us,' he shouted, dizzy with the absurdity of it all.

Abu Laith opened the window, and the fresh cold air streamed in. 'Welcome,' he cried to the street. 'Come in and see how the Daeshis live.' He thought about the neighbours who had denounced him to Daesh, and how afraid they would be now. He felt his time had come again.

There was a pile of make-up by the mirror, and Abu Laith rifled through it, opening the pots. He found a pink lipstick and screwed it open. Carefully, he wrote on the mirror.

'This is the place where the pimp imam lived.'

He signed off with a flourish. 'This is the man who was decreeing what was *haram* and *halal*,' he roared. 'And he was the most *haram* himself.' He strode into the garden and out to the group at the front, throwing handfuls of the underwear like confetti. On the road, everyone was laughing.

Abu Laith pranced about holding bras up to his chest. 'Let this be a reminder for everyone,' he said through his tears of laughter. 'That if Daesh comes back we all have to start wearing things like this.' One by one, he hung the bras – pink, purple, red, white, blue – on the bars that covered the outside of the window facing the road.

Back at his own house, Abu Laith greeted neighbours he hadn't seen in years. Most of them had thought he was dead, or that he had abandoned his family in Mosul.

'Praise be to God, you survived,' said one of his visitors. Abu Laith revelled in it. He was an important man again. Everyone knew he had never joined Daesh. Of the neighbours who had denounced him and sucked up to the Daeshis in the mosque, there was no sign. All day, he held court, organized teams to sweep up the rubble and ran in and out of the zoo to check on Zombie and Lula.

Around 4 p.m., one of the army's Humvees rolled into the street and stopped outside the house. Soldiers started getting out. Abu Laith was about to greet them when a smaller figure stepped out, straw-blonde hair hanging loose and wearing a wide grin. For once, Abu Laith couldn't say anything. He was crying, holding his daughter to his chest.

With the family gathered around her, the children pulling at her uniform and laughing hysterically at their luck, Dalal was borne inside the house. Geggo, Mo'men and Shuja, who didn't remember their older sister, were wailing. They all sat down on the sofa together. Abu Laith wouldn't let go of her hand.

That night, someone brought a speaker onto a nearby street. Soldiers poured in from their bases and from around the neighbourhood; locals ran over to see what was happening. On the street near Abu Laith's house, screaming with delight, the crowd jumped together, up and down, the soldiers shooting in the air, as they partied for the first time in two and a half years.

38

Abu Laith

THERE WAS SOMETHING FUNNY GOING ON WITH THAT LION, Abu Laith thought, as he took his customary turn around the zoo one morning. It was still cold, but the winter sun was pouring through the bars of the cages. They were still encrusted with bits of old debris and food that no one had been able to work up the courage to clean out, with the lions as hungry as they were. The family were all tired after the liberation. There still wasn't nearly enough food, and they had no money to buy the animals the meat and vegetables they needed, to Abu Laith's shame.

The animals deserved more, he thought, particularly Lula. A few hours after Warda's remains had been taken away she started looking for him, frantically searching each corner of the cage, batting aside the plastic bucket as if hoping to find him hiding behind it.

Abu Laith had become deeply aware of the world's unfairness. Warda was good, and deserved to be alive. Lula deserved to have her child. Her mate, the male bear who had been left behind in the old zoo by the Forest, had been shot by Daesh because he lived right by their training camp. Each time Lula woke up in the morning now, she started her hunt again, scraping at the bars for a trace of Warda's smell, howling when she didn't find him.

Abu Laith wanted to give her honey and apples. Instead he usually had only a pan of old bulgur, or some half-rotten vegetables. When the electricity had been shut off, and the generator ran low on fuel, Lumia had wanted to eat all their meat before it went rotten. But as the fighting worsened, they hadn't been able to cook. By the time the shelling had calmed enough so that Lumia could make it into the kitchen, the meat was turning green and inedible. Zombie, Abu Laith thought, as he took the meat to the lion, wouldn't mind.

Zombie's parents were a different matter. Mother had been acting up for days, roaring at passers-by and pushing Father out of the way when they were fed. As far as the children were concerned, this was of a piece with her general level of cruelty. Mother had been in disgrace since she had eaten Warda's arm. Though they maintained an outer veneer of civility so that Abu Laith wouldn't shout at them, they thought of her as little more than a murderer. Their ill-feeling was compounded by Warda's death and by Mother's disgraceful attitude towards her mate and child.

Father was not well. Not too long ago, his muscles had rippled under his sandy hide. Now his bones were just the frame for draping skin – dried crust-yellow and strangely thin. For an entire day, he had been lying still, deep in sleep and occasionally stirring in his dreams.

The lion's condition was, Abu Laith knew, serious. He had tried every trick he knew to wake him up. Vegetables and loud encouragement had no effect, and the lion was still sleeping. Mother, on the other hand, looked rangy, but nowhere near emaciated. It was not hard to figure out why. Try as they might to feed them separately, Mother always stole Father's food.

It was the same today, as the children did the morning round of the zoo with Abu Laith.

'I wish we had another cage,' moaned Abdulrahman to one of the younger kids, who was engaged in a dangerous game with Mother. Crouched on top of the cage like a monkey, he hung his arm down, limply, to tempt her to jump. As she crouched, ready to leap at this enticing prey, he pulled it up and laughed.

Abdulrahman shoved a bit of bread at Father, who was listless as ever. 'Take it,' he said. 'She's not looking.'

But she was quicker than he'd thought. Smelling the food, Mother abandoned her twigletty human meal to pounce on the bread. She picked it up in her mouth and dropped it. Father didn't want it either.

In the cage next door, Zombie lay quiet. He was more gaunt than Mother, but he could still walk. The other day he had eaten an apple that one of the children had offered him as a joke. He must, they thought, be pretty desperate. But there just wasn't enough food for them, with prices still incredibly high and shops and suppliers destroyed by the war.

'I wish we could put Father in Zombie's cage,' said Abdulrahman, not for the first time.

Abu Laith had told them it was impossible. Two male lions in the same cage would attack each other, he said. But as they watched Mother bat away the bread, ignoring her mate, they all knew that this would not end well.

The next day, Abu Laith's morning round started as usual. Abdulrahman walked next to him, intensely serious, the other children following at a respectable distance. First, as always, they checked the poles that held the gate in place, giving them a solid kick each, and shaking them to make sure they were anchored.

Once Abu Laith was satisfied, they looked in on the squirrel, who was bouncing around his mesh cage. At someone's insistence a few years ago, funds had been procured for a hamster wheel, which spun constantly as the squirrel took his exercise.

Abdulrahman checked the water and food, as he had been taught, and they walked down the path towards the pelican pond. On the way, they passed two of the carousels that had survived the airstrike. 'Horrible things,' growled Abu Laith, as he did almost every time he saw them.

At the pelican pond, there was a moment's contemplation at the water's edge. The bird had been one of the casualties of the airstrike. Abu Laith thought the shock wave must have killed him. Pondering this, Abu Laith and Abdulrahman crossed the bridge – an architecturally redundant 15-foot high-arched walkway – over the water.

Through the trees they could see the lion cages. Mother was standing up, crouching over a large form on the floor of her cage. Her head was down, and she seemed to be eating.

'What's she doing?' asked Abdulrahman.

But Abu Laith was already running towards the cages, shouting, as the lion tore and chewed at the ruined remains of what had yesterday been her husband.

As Abu Laith drew close, Mother lifted her blooded mouth. Father's ribs were open beneath her, skin shredded back in strips. She tore into his belly, through muscle and viscera and tendon, blood thick on the floor, gorging herself.

Abu Laith threw himself onto the bars, and she yowled like an angry cat. 'Move,' he shouted. 'Get off him.'

But there was nothing he could do. Shouting and cursing the day the lion was born, Abu Laith stood by the cage, helpless with fury. He knew that she was an animal, with an animal's instincts. But this was cruel.

Abdulrahman stopped in front of the cage and tried very hard not to be sick. Mother tore off a strip of her mate and chewed. Behind him, another half-dozen children were shouting and bawling.

Abdulrahman felt like he should say something. 'It's because she's hungry,' he shouted over the sounds of ripping and tearing behind him. 'If they'd had enough food, she wouldn't be eating him.'

The other children disagreed. 'She's evil,' shouted one of them. 'She killed him and ate him.'

Abu Laith, who would usually have been inclined to box his ears, sank back feeling defeated.

39

Abu Laith

Marwan came back about a month after the liberation, bone-tired and thin as a ferret with a dodgy new haircut that immediately earned a reprimand from Abu Laith. He had seen some of the worst of the fighting in the east, hiding out for weeks with his family before escaping through the lines to a displacement camp outside Mosul, where he'd lived in a tent in a sea of mud. Sick and hungry, he'd walked back to Abu Laith's house, only just managing to stay conscious.

When he arrived, he'd barely said anything. He ate half a helping of rice and fell asleep, almost immediately, right there on the floor of the living room. The children picked over him like birds, until Lumia chased them away. He didn't stir for a whole day, and when he awoke, he was still exhausted.

The east had been bad, he said, but not as bad as the west. He'd met people from there in the camp. Some were so thin they could barely walk, little more than skin-covered bones. The Old City had been all but destroyed. Abu Laith couldn't quite believe it. Those lanes and alleys were the heart of Mosul, which kept the city together.

'There's rubble everywhere,' Marwan said. After a while he asked 'How are the animals?'

Abu Laith told Marwan about the airstrike, and the baboon, and Warda.

'And Heba?' Marwan asked, after he had finished. 'My fiancee. Has she come?'

Abu Laith had been expecting this, but it was still difficult. He had offered to pay for their wedding, knowing that Marwan's own father wouldn't. Since Marwan left, he had kept an eye out for a young woman by herself, looking around the animal enclosure. But he had no idea what she looked like, and she would never know to ask him about Marwan.

'No,' he said. 'She hasn't come.'

Marwan nodded.

Abu Laith knew he was devastated. 'Come on,' he said. 'We have a lot of work to do.'

They spent the afternoon clearing out the zoo, throwing buckets of water over the cages, though it didn't seem to make a lot of difference. The dirt had been baked in by the heat, and they couldn't get inside to clean them properly. Mother tried to attack anyone that went near her cage, and Lula – who was obviously traumatized – shied against the bars if anyone as much as moved near her. Marwan, who was tiring easily, took breaks in the shade of the blood tree, stained dark where the Daeshi had been pinned up like a mannequin. He could barely lift a shovel, but he carried on because it felt good to be doing anything at all.

The children were sent on a foraging mission and returned back some hours later scratched, flea-bitten and very pleased with themselves after an afternoon in the ditches and dumps of Mosul. They carried a bag of vegetable peelings, off-cuts of meat and old rice. Zombie tore into these scraps with glee, but Mother approached the food warily. She had, Abu Laith decided, lost her mind.

As she sniffed at the scraps, Marwan stuck a broom through the bars of her cage and dragged out the mess of dried sinew and bone that remained of Father. With the winter sun burning the back of their necks, they dug a shallow hole in front of the cage. There was little ceremony; within a few minutes, Father's remains were interred in front of the cage where his cannibal wife reigned, and where he had spent the last years of his eventful life.

The next morning, Marwan had taken up his usual place as custodian of the animal enclosure. Abu Laith was still at home, making plans to find new sources of food. The meat was too expensive nearby – prices many times higher than they were before Isis – and the lions needed at least 10 kilos a day if they were going to thrive. Lula needed honey and apples.

It seemed almost impossible, thought Marwan, that the animals would ever be able to survive. Daesh had gone, but most families in Mosul were still living off rice and bread. Whenever he looked at Mother, yowling in her cage, and Lula, who was still obsessively looking for her cub, he didn't know how much more of this they could take.

His thoughts were interrupted by a loud clanging on the gate, which he had kept locked on Abu Laith's instructions. He, in turn, had been told to keep it shut by the army because of the piles of ammunition and mortars still stored under the carousel.

Marwan walked over towards the gate. Someone was shouting, trying to shove it open.

'Assalamu aleikum,' called Marwan as he approached the gate. 'We're closed.'

'You open this door right now,' barked a voice behind the gate. 'I'm from the police. Let me in.'

Marwan could hear muted giggling when the man stopped talking. Intrigued, he opened the gate and stuck his head out.

A skinny, indignant little man with a thick moustache was standing in front of the gate. He was wearing the blue camouflage uniform of the federal police forces, which bagged over his coat-hanger frame. Behind him were two girls in their twenties, dressed in jeans, long shirts and hijabs. They looked at Marwan and giggled again.

The policeman did not waste time on introductions, and tried to barge past Marwan into the zoo. Thin and bedraggled though he was, Marwan stopped the man easily. Grabbing the front of his uniform, he shoved him back into the road.

'We're closed,' he said, and the girls giggled again.

The policeman looked furious. Clearly, he had wanted to impress the girls by taking them to the zoo. But now he was being stopped by a younger man who was treating him with open disrespect.

'I have an order from the Golden Division,' shouted the man. This was the elite of the Iraqi army, venerated for the way they had worked through the country destroying Daesh. Now they were fighting in west Mosul; certainly, Marwan thought, not spending their time issuing policemen with permission to go to the zoo.

'No,' said Marwan. 'You're not coming in.'

The policeman launched himself at the gate. 'You're disobeying a direct order,' he shouted. 'I'll have you thrown in prison.'

'Go and get Abu Laith,' Marwan shouted at one of the children, who was loitering around, enjoying the spectacle. 'Tell him to come now.'

The boy ran off.

'They'll let me in,' said the policeman to the girls. 'These villagers are so suspicious.'

A bark of fury signalled Abu Laith's arrival. Red-faced and purposeful, he strode up to the zoo gates. 'Who are you?' he shouted at the policeman.

'I'm from the federal police,' he said. 'And I order you to let me in the zoo.'

'Can't you read?' Abu Laith spat, indicating the sign he had put up a few weeks before. It read 'no entry' in his spidery script.

'Of course I can read,' said the policeman. 'What have you got in there that you're hiding?'

'That's between me and the army,' Abu Laith later remembered saying. 'They left me in charge. There are important weapons here, as well as the animals.'

'I've heard there's a lion in there,' smirked the policeman. 'I'll shoot it in the mouth. Then it'll stop roaring.'

He hadn't finished the sentence before he was on the ground, and Abu Laith was pounding his head into the concrete, shouting incoherently. 'You filth,' screamed Abu Laith, finding words as he pummelled the policeman with his great arms. 'You coward.'

Marwan watched, fascinated. The old man had pinned down the younger one's arms with his knees, fists smashing into his face with a rigour and tempo that Marwan, who thought of himself as a seasoned brawler, could only envy. The policeman had already stopped resisting, and all they could hear was the wet thwack of Abu Laith's punches, and the girls' screams as they watched from a safe distance. A few of the neighbours ran over, looking concerned.

'That's enough,' said one of them, sternly. 'You're going to kill him.'

'That's right, I am,' shouted Abu Laith.

'OK, fine,' said the neighbour. 'But the rest of them will come and kill you if you do.'

Abu Laith ignored him. Marwan, however, saw his point. He grabbed Abu Laith and forced him up. Pushing him to the side, he bent down to look at the policeman. His face was a mess of blood. But he was conscious, and breathing.

Abu Laith leaned over him. 'You're banned from the zoo,' he said. 'And don't bother coming back. I'm staying here until you murderers leave. And if one of you tries to touch the animals again, I'll break your legs.'

Glancing around him like a haunted man, the policeman stood up and hobbled away towards the road, the girls following after him.

'Nice punch, Abu Laith,' said Marwan.

40

Hakam

Isis hadn't been the worst houseguests, all told. There were empty cans of Red Bull and a few pans of half-rotten beans scattered around the kitchen, and some of the windows were smashed. But the Daeshis who had used the Zararis' house as a rocket emplacement had left behind much of the food and even one of Arwa's rings that she had forgotten on the counter when they ran. It was lying on the ground in the corridor, which was scattered with broken plaster and debris.

Even the garden hadn't fared too badly. 'My babies,' laughed Hasna, as she ran out to the chicken coop, greeted by clucking. 'You're alive.'

Two of the chickens had been stolen – though the thief had considerately shut the coop door after him – and the birds of paradise were gone. Some of the rare roses had been pecked to pieces by stray chickens from the surrounding houses, turfed out by their owners when they were forced to leave their homes. The Zarari chickens had, at least, been left with food and a plentiful water supply.

The roof was littered with bullet casings. When one of Hakam's uncles had gone to check on the house before the family returned, he had found five rocket-propelled grenade launchers under the parapet. He had carried them out on the street and left them there.

'Good thing you did,' a soldier cheerfully informed him when he jumped out of a Humvee to pick them up. 'Otherwise we'd have burned the place down.'

As it was, nothing was burnt, but plenty was broken. None of it mattered that first day. The family were dirty, exhausted and emaciated, but they were home, and they were alive.

'Hassan,' shouted Arwa, as her eldest son's face lit up the phone screen. 'We're OK.'

The family waved down the phone to their brother in America, where it was late, and he was crying with relief. There had barely been any signal for weeks, and Hassan had been terrified that the worst had happened. From him, they learned that their extended family had survived. In Mosul, the signal was so bad that they could almost never get hold of each other. But through Hassan, sitting in Pennsylvania, they realized the others had made it.

The clean-up took almost two weeks, the family on their hands and knees as they worked to purge Daesh from the house. Arwa swept the broken glass from the floor, and taped up the windows. Said, going through his wardrobe, found that some of his clothes were missing. He thought that the Daeshis might have used them to blend into the civilian masses fleeing from the fighting.

The lock on Hakam's door was broken, presumably by looters. Inside, his clothes were strewn about. Someone had stolen his watches, and used up almost all his cologne. But his guitar was still there, protected under the prayer rug.

He picked the guitar up and sat down. For the first time in

a long time, his fingers twitched over the familiar strings, and he began to play.

It didn't take long before everything looked almost normal again. From morning to evening, the family had tidied the house, throwing out the rubbish of occupation. The army was now milling about on their street, sometimes coming in to borrow water from their well. The family slept in their beds on clean sheets.

But Hakam couldn't relax. At first, he had tried to stop Said digging up the books. The army felt alien, temporary, and he couldn't shake the feeling that Daesh would come back.

His father, backed by Hasna, ignored him, and from the rich earth under the orange trees the well-thumbed texts on Islamic jurisprudence – and Miles Copeland – were dug out to take pride of place, once more, on the dark wooden shelves in the living room. Out, too, came a SIM card, buried when Isis had announced that owning a phone meant death, and all of Arwa's gold.

But everything was still wrong, somehow. Hakam couldn't be happy – his uncle had been killed, and his city lay in ruins. Yet he had survived, so it felt ridiculous to sit around grieving. Hasna felt the same unease. Nearly three years had been taken from her. She couldn't get back to normal. Even reading felt strange. Nothing felt safe.

About a week after they got home, Hakam had turned his phone on and logged into Facebook for the first time in almost a year. He had hundreds of notifications. Friends, girls, photo tags, protein powder, football and Daesh. A lot of them were invitations to play online games.

His timeline was filled with people who had spent the last two years living with some degree of liberty. Taking an entire album of selfies, sending hotdog legs from the beach, smoking hookah on nights out. Messages had started frantic, tailed off and started again in the last few days, as the liberation of Mosul dominated the news.

As he stared at his timeline, a feeling of heaviness crept over him. He didn't want to talk to anyone. He re-activated his messaging apps, without realizing it would send a notification to all his contacts. The texts kept pouring in, from the US to Germany, clamouring for attention. He knew they meant well, but he was overwhelmed, and didn't know what to do.

Forcing himself to type, he sent a couple of messages to his closest friends, telling them he was alive. When it was done, he turned the phone off and fell into a nightmarish half-sleep.

A couple of weeks after they returned home, when the house was finished and the mortars crashed less often, he was feeling a bit better. He had eaten, showered and spoken to his aunt in Erbil on the phone. Her joyful delirium had shaken him up, and he began to feel he could go back to normal again.

He logged on to Facebook, replying to messages and tag requests. His friends had missed him, and they were overjoyed that he was alive. He was about to put his phone down when he saw that someone had posted a picture of a half-dead bear, next to a half-dead lion. They were filthy, and looked little more than skin and bones. He had to look closely to make sure the lion wasn't a dog.

'This is how the animals are living in Mosul Zoo,' read the caption, in Arabic. 'Can anyone help?'

Hakam was intrigued. Looking closely at the picture, he saw that it must have been taken round the corner from his house, in the old park. He had no idea there was a zoo there, though he knew there was an amusement park that had been open under Isis.

In the comments section a charity had written, in English, that they could help. Hakam thought for a moment, and commented: 'That zoo is by my house. Let me know if I can do anything.'

Minutes later, they messaged him. He grabbed his phone, stood up and went for a trip to the zoo.

41

Abu Laith

Abu Laith had been right. Night passed and morning came, fresh as spring, and the police left the zoo alone. No one had seen the skinny officer since that afternoon.

As he had every night since the policeman had come, Abu Laith sat by Zombie's cage, crooning to the lion. He needed to protect the animals from the thieves who were scattering around the streets of Mosul, picking up what they could in the chaos. It had been a free-for-all since the army came, and people were guarding their families and their belongings with their lives.

Abu Laith was no different. Each day, he or Marwan stood guard in the zoo. Each evening, Lumia brought Abu Laith out an old blue blanket and two pillows so that he could bed down for the night. When she made dinner, she'd send Abdulrahman out with a portion for her husband, with dire warnings of what would happen if he got himself shot while protecting the lions. Abu Laith would spent the nights on high alert until, when his eyes gave in, he would bed down next to the cages.

It didn't take long before his perseverance paid off. Around

10 a.m. one day not long after liberation, a group of soldiers led by a burly commander came in two cars, a civilian in tow.

Abu Laith had been at home, and Marwan had let them into the zoo. They had said they were there to inspect the ammunitions cache Daesh had left behind, and Marwan showed them where it was.

They'd taken a cursory glance at the weapons hidden under the carousels before asking to see the animals. Marwan, suspicions growing, led them over to the lion cages. He'd skulked in the corner while the civilian, who had kept quiet until now, looked through the bars, inspecting the animals with a practised eye.

As Marwan stood nearby, the group switched into Kurdish, presuming that the apprentice zookeeper wouldn't understand. What they didn't know was that Marwan had grown up speaking Kurdish, and knew perfectly well what they were saying.

'What do you think, could we sell them for $3,000?' the commander asked, as the civilian peeked into Lula's cage.

'Maybe a bit less,' the civilian, who was wearing a suit, said. 'I'm not really sure if they'll survive.'

The commander hummed and hawed for a while, as the civilian wrote down a few notes. 'I'll give Duhok zoo a call,' he said. 'They might take them.'

'Tell them we can pick them up tomorrow,' the commander said. 'And that we want the full amount in cash, right away.'

They made to leave, thanking Marwan in Arabic, and he followed them. When they neared the gate, he spoke to them in Kurdish. 'You know you have to ask permission before you do that kind of thing?' he asked, nonchalantly. 'You can't just go around selling other people's animals.'

It was the commander's turn to gape. Marwan signalled for one of the children to go and find Abu Laith. 'Are you a vet?' he asked the civilian. 'Because if you are, you really should be helping animals, rather than stealing them.'

The conversation was cut off at that point by the arrival of Abu Laith, motoring down the road at top speed.

'These people are trying to steal the animals,' Marwan shouted as he drew closer. 'They were talking about it in Kurdish, but I understood them.'

Abu Laith, who had not liked the cut of their jib, immediately understood the gravity of the situation. 'Over my dead body,' he roared.

'It's for security reasons,' the commander said, changing his tack. Abu Laith had to be restrained by the group of onlookers that had gathered around them. A standoff ensued, the would-be thieves insisting that they were permitted to drug the animals and take them away.

Twenty minutes later it ended after the commander gave up, slightly spooked after Abu Laith called a friend of his in the Christian town of Hamdaniya who – he assured the Kurdish vet – had connections with the intelligence services.

Abu Laith hadn't budged from his sentry post outside Zombie's cage since. Wrapped in his old blanket, staring up at the stars with a lion beside him, he felt truly at home. 'Soon I'll get you a live goat to hunt,' he would promise Zombie, as they sat under the night sky. 'You'll need to train your instincts if you're going to be able to survive in the jungle.'

Abu Laith believed, as he always had, that one day Zombie and Lula would be free – ideally close enough that he could visit them – but at liberty to run and roam and kill. They would hunt animals like the ones on the National Geographic channel: in the forests or across the savannah.

'In the wild, you'll see a goat, or a sheep, and when you do, you have to get right down low,' Abu Laith would say. 'You have to stay right down, or they'll see you. Sniff the wind' – he would sniff, looking about – 'and know whether your prey will smell you. If they do, they'll run. You don't want that. Then, when you get close enough that you can see its eyes shining, you take a deep breath, you draw yourself together, and pounce. You'll get the idea.'

The lion never responded, but Abu Laith thought he understood. 'Soon,' Abu Laith would say. 'You'll have a live goat to hunt, and I'll have whisky.' With these thoughts in mind, man and beast would sleep through the night, dreaming of hunts on the endless savannah.

The next few days passed slowly, the hunt for food as desperate as ever. Abu Laith and the children veered between army checkpoints in the few streets they could freely navigate. Security was tight, the fear of Daesh attacks ever-present.

In the courtyard, the children shoved each other and held spitting competitions – just as bored as they'd been under Daesh. Lumia was still too afraid to go outside.

Holding a brace of buckets, Abu Laith trudged up the road from the zoo to the local shop. The day before, he'd gone to the illegal sheep market out by the city walls – a dusty moonscape marked out by its squat brick buildings and the metal-rank dead smell of blood and offal. But there was never enough money, and the shopkeepers were tired and angry at the pushy man from the zoo, who always wanted more.

They could still hear the fighting, though it was further away in the west, where Daesh was holding out in the rat-run alleyways of the Old City, thousands of civilians taken as human shields around them. Almost every day Abu Laith heard the distinct boom, not from a plane or a mortar but the explosion of a *mufakaka*, a car bomb plated in thick sheets of metal, launched at soldiers with a man inside.

But the area around the zoo was, now, relatively safe. He peered around the street corner towards the army checkpoint set up by the edge of the zoo a few weeks ago. There was nothing there – just a few soldiers loafing around a house that they'd requisitioned as a temporary base.

He was about to cross the road to the shop when he saw a tall young man with glasses. He was walking along the zoo fence, peering inside.

The young man saw Abu Laith looking, and smiled. 'Assalamu aleikum,' he said.

'Wa aleikum assalam,' replied Abu Laith. He liked this young man, who seemed calmer, somehow, than everyone around them. He was dressed like an American, in a t-shirt and jeans, and his skin was pale as porcelain, but when he spoke there was no doubt he was a Moslawi.

'My name is Hakam Zarari,' said the young man, stretching out his hand. Abu Laith put down one of the buckets and shook it heartily.

'Abu Laith,' he said. 'Welcome. Did you come to see my zoo?'

42

Dr Amir

Dr Amir was in the office, a place where he never really felt at ease, restless under the strip-lighting. Everything was scrupulously clean, the birch-wood desks all in order. The coffee he had made in the office kitchen sat on his desk as he scrolled through his overflowing inbox. He drank many cups a day, and this was not the first.

It was a freezing January day in 2017, and the streets of Vienna were windblown and brown, quiet but for the harried office workers streaming down the street. It was at times like this that he dreamed of Cairo – the hot energy of that vast city. For thirty years, since he graduated from vet school, Dr Amir had lived in Austria, amid the palaces and corniced cafes that once were on the very edge of the Occident.

He was, he liked to say, a stray dog. Jocular and relentlessly comfortable, with a steady grin and a crinkled forehead topped with a block of black curls, he was likeable in the extreme, his restlessness betrayed only by a certain ranginess of gait. Through stoic patience and a gentle manner that masked unrelenting determination, he – along with his colleagues

– had created an organization that was unlike any other. For twenty years, Dr Amir had been the world's foremost conflict rescue vet, tracking down and helping animals living in dangerous conditions across the world.

It had begun when he was at university in Vienna, where his mother had sent him to supervise his sister, who had just married an Austrian-Egyptian man. After a few days in the Austrian capital, the brother-in-law was declared fit for purpose, but Dr Amir decided to stay anyway.

He was twenty-four, and he had moved from the chaotic dust-caked streets of Cairo to a country where jaywalking was frowned upon. His main skill – an in-depth knowledge of the physiological intricacies of cows and water buffalo, expressed exclusively in Arabic – was not of much use. Settling down, he dutifully learned the German words for dog (*Hund*), cat (*Katt*), bovine kidney stones (*Rindernierensteine*), and got to work.

In Austria, he found, he could learn about wild animals, with their instincts and inner laws that could never be overridden. They killed because that's what they had to do, not because of anger. Unlike humans, they didn't change. He had wanted to be a vet since he had first sat in front of the family TV in Egypt and watched Daktari – a show about a square-jawed hero who rescued animals in the jungle in Africa. Now he knew how to achieve it.

One day at university, where he was studying for a masters, he saw a poster from a charity called Four Paws (*vier Pfoten*) bearing the legend: 'Student needed to castrate dogs in Romania.' Amir was encouraged, and wrote the address down. Soon after, he presented himself at the organization's headquarters in a stately building on Mariahilferstrasse.

'I would like to apply for the dog castration position,' said Dr Amir, as he sat down with a young volunteer named Marian.

'Sure,' she said.

Within a few months, Dr Amir was tearing around the Romanian countryside in a black Mercedes van, knife in hand, his wrist action now so deft that he could emasculate eighty dogs a day. In his spare time, he waged a gentle war on the Romanian government, cajoling them to force through a bill that would ban the killing of street dogs – which were seen as a menace – and institute a state programme of castration.

The patriarchal forces that governed Romania found the idea repellent. To emasculate a dog, they explained, was to ruin its life and interfere with the will of God. Dr Amir wore them down, learning Romanian so that he could communicate with the locals and press his case.

They eventually agreed to his proposal, and he returned triumphant, having organized the innovatively named First International Stray Dog Conference in Romania. The next year, he travelled to Bulgaria to convince the Roma population to give up their dancing bears.

After a few weeks of investigation, travelling around their encampments and the trains and town squares where they performed, Dr Amir realized that the 'dancing' was nothing more than a Pavlovian response. When they were cubs, the bears were taken from their mothers and forced to stand on a hot plate, their paws smeared with Vaseline to protect them from burning. As they lifted each paw in turn to avoid the heat, the Roma trainer would play music for them. In time, the sound of a violin or a guitar would make them jump around. But bears, he explained decades later, were not musical animals. They could smell well, but not hear well. The dancing was torture for them.

Dr Amir was appalled. With the determination of a zealot, he pressed flesh in government offices and haggled with scarf-draped ladies in remote encampments. The government passed a law, and the dancing bears no longer entertained commuters on the Bucharest metro.

Life entered a certain rhythm. As word spread about the wily Egyptian vet, the work came rushing in. Following a tip from a concerned visitor to a Black Sea resort in Romania, Dr Amir and an assistant travelled to a beach town where three drugged lions were being kept in a nightclub – rented out as photo props to holidaymakers for a few coins each evening – and in a petting zoo at a skiing resort.

Disguising himself as an Arab sheikh, Dr Amir set up shop in a casino, drinking coffee to stay awake until at last he met the lion's owner. Putting on a thick Egyptian accent, he told him that he wanted to buy the lions for his Romanian girlfriend, and offered a good price. The man agreed. The police pounced, and the lions were rescued.

As he ricocheted around the world saving animals from various catastrophes, Dr Amir began to see the world in a different way. Wherever humans suffered, animals suffered too. Their food, their care and their lives were almost always the first casualties of war: pet dogs left to starve because the owners had to feed their children, zoo animals abandoned when keepers fled from fighting.

He thought back to his childhood in Egypt, and how the other children used to throw stones at dogs. Whenever he'd try to stop them, they would beat him up. For Dr Amir, hurting animals had always felt wrong, just like hurting humans.

As he watched the news, and the terrible suffering it chronicled, he thought of the animals who died among humans, and at their hands. Someone, he thought, should care for them too. If people cared for animals, they should care for humans, and if they cared for humans, they should care for animals. Kindness should not be divided.

In twenty-five years of careful organization, failure and exhilarating success with Four Paws, thousands of animals had been rescued and thousands more born from them. Humans were being trained to think differently about the animals that shared their world.

The walls of his office were lined with books, two deep in places, of rare diseases that affected animals from Sweden to Sumatra. Bears, flamingos, giraffes and tigers all fell within his area of expertise, as did the predatory pets of Middle Eastern dictators – a subject with which he was intimately familiar. Uday Hussein's nine lions, abandoned in his father's palace grounds after the Americans swept into Baghdad. He searched for, but never met, Saif Gaddafi's two white tigers, who had – for a time – also lived in Vienna when the colonel's son studied there, passing their days in luxury at the Schoenbrunn Zoo, their bills presumably paid from Libya's oil budget. Instead he found monkeys, lions, hyenas and deer in Tripoli that had been abandoned by Moammar al-Gaddafi when he fled. They were saved. Others fell by the wayside.

One ex-wife and three daughters – who had grown up in Vienna as he hacked through jungles and dodged bullets – could testify to Dr Amir's love for his job, and occasional blindness to all else. The flat he now owned in the 22nd district of the city stood mostly empty, suitcases lining the hallway and clothes rarely put away.

The little time he spent in Vienna, he was in the office – now a cream-coloured three-storey building off the Linke Weinzeile, a distinctly unlovely stretch of dual carriageway far from the Mitteleuropa grandeur of the city centre. He shared an office with Sabine, Marlies, Noha and Marion, upstanding ladies devoted to the doctor's mission. They all thought he was mad, but it was a madness they admired. When he started on a project, the doctor's fierce single-mindedness had a way of sucking in everyone around him.

Leaning against the window in the corner where Dr Amir sat was a picture of him – his portly face beaming – alongside Brigitte Bardot, langorous eyelids half-shut in the camera flash. She had flown to Romania to congratulate him when she'd heard how he had saved the stray dogs. She later financed half of the dancing-bear project in Bulgaria. Next to it was a photo of two golden-skinned little girls. He had been looking after two of his daughters one day when an emergency with the dancing bears called him to Bulgaria. They'd had to come. He thought they'd had a nice time.

On his laptop screen was an open Facebook conversation from his most recent case, one that the staff at Four Paws had stumbled across online. Trawling through social media posts about Mosul, which was in the midst of a bloody campaign of liberation by the Iraqi army, they had come across a message posted by a resident of the city living in an area just cleared of Isis fighters. 'Does anyone know what to feed lions?' she had written. 'There is a zoo near my house and the animals are dying.'

He jumped between the tabs on his laptop. One had a map showing the location of the zoo relative to the front line, a line that Dr Amir knew was moving every day. Another showed a picture of a bear, asleep or unconscious on the floor of her filthy cage. The last showed a lion, mane matted into a sienna dreadlock.

'What should we do?' the concerned Mosul resident had asked.

'Leave it to me,' one of Amir's colleagues had replied on Facebook, and passed the case on to the doctor.

He clicked back on to the photos. The lion, he thought, had a clear case of malnutrition. He couldn't see a water bucket in the photo, and no food. The prognosis was, the doctor concluded, very bad.

They would have to act. He sent a message to Dr Suleyman, a distinguished veterinarian and old friend living in Erbil, a Kurdish city 50 miles from Mosul. 'I'm coming to visit,' he said.

Dr Suleyman, who was in his fifties and living comfortably, was game. They had met five years previously, when Dr Suleyman had contacted Four Paws for support. He had, on his own initiative, begun a programme to castrate the stray dogs of northern Iraq, and came across Dr Amir's extensive work on the subject in Bulgaria.

The two had finally met at a conference in Cairo where Amir, taken by the Kurdish doctor's enthusiasm, encouraged him to attend a training session in Bulgaria. The course was a success, and the stray dogs of northern Iraq were soon much less fertile. Half a decade later Dr Suleyman, like a lot of people, owed the doctor a favour.

Thankfully, Dr Amir reflected, so did a lot of Iraqi officials. In the days after the American invasion of Iraq in 2003, the doctor had taken it upon himself to rescue the animals of

Baghdad Zoo, whose keepers had abandoned the premises when the Americans swept in.

Of the 720 animals that had lived there before the invasion, only twenty-seven remained by the time Dr Amir made it through the blocked roads with an Iraqi refugee vet he had picked up in Vienna as a guide. The refugee's family, who had thought he was dead, were happy to have him back, and – as was customary – also owed Dr Amir a lot of favours.

The zoo was ruined. In the chaos and looting that followed the withdrawal of Saddam's troops, someone had stolen the giraffe, and many of the other animals had been killed for food. Even the aquariums had gone. Lions and dogs were locked in the same cages, with predictably bloody consequences.

Days before the doctor arrived in Baghdad, another animal in the zoo had died – a Bengal tiger shot in the head by the American soldiers who had turned the park into their base. It later transpired that the killing was an act of revenge by the Americans. A group of infantrymen had been drinking, and one of them had put his hand inside the tiger's cage. The animal had, of course, swiped at it, biting a finger off. The remaining soldiers took up their M16s and unloaded into her head. To his eternal regret, the doctor had arrived too late to save her.

The new case surely wouldn't be that bad.

His colleagues in the office did not agree. 'What are you thinking?' Noha had asked. She told him it wasn't safe, and that the area had only been liberated a few days before. There was still fighting going on.

Marlies handed him a folder of information gleaned from open-source reporting of the conflict that they had prepared for him. She told him that the foreign ministry had asked her, in all seriousness, what was wrong with Dr Amir.

He had not been impressed. 'I have to go,' he said, and began preparations to leave.

For all of his occasional bouts of dreaming, the doctor was a man of action. Within a few days he had charmed several Iraqi politicians and bought a flight to Erbil, leaving in two weeks. Amid the carnage of war, the doctor would launch a rescue campaign.

As he made his preparations to leave, Dr Amir's phone pinged with a Facebook notification from a name he didn't recognize. It was an answer to the request he had left on the woman's status about the bear and the lion.

'I live around the corner,' wrote a young man called Hakam Zarari, whose profile picture showed a thin face with thick-rimmed black glasses. 'I can help.'

Dr Amir typed out a reply.

The two men chatted for a few minutes, each struck with the goodness of the other. Quickly, a deal was made: Hakam would go to the zoo and see what the animals needed, sending pictures and videos to Four Paws. From there, the vet could make a diagnosis and provide a plan.

Hakam, Dr Amir surmised, wanted to do something good. As a young man in Egypt he'd known something of the complete and crippling boredom that can strike in difficult situations.

As Hakam signed off, Dr Amir looked at the pictures again. There was no doubt that the lion wouldn't survive for much longer – maybe just days from when the photo had been taken. They could only hope, he thought, that it wasn't already too late.

43

Dr Amir

The morning of 23 February had broken, fresh and grey, when Dr Amir finally set out for Mosul from Erbil. It had taken him two weeks to get to this stage, bullying and cajoling a raft of soldiers and Iraqi officials past the point of sanity.

Next to him in the car sat Dr Suleyman, who was deeply concerned. There had been a certain reluctance on his part to take part in this mission. He suspected that Dr Amir was up to something, and he was absolutely right.

Dr Amir had told his friend that they would only be going out to the edge of Kurdish territory, 20 miles from Erbil, to look at Mosul from a distance. This was not true. Dr Amir had every intention of going into Mosul that day, and bringing his Kurdish colleague with him, despite rumours that the border was closed.

The doctor wasn't particularly proud of his deception, but it felt necessary. While Dr Suleyman was a relatively adventurous man, and very fond of his Egyptian colleague, he had several children and a wife who would never let him risk his life to go to Mosul. But Dr Amir didn't know many other Kurdish people who could go with him, and help him through the

Peshmerga checkpoints. The decision had been made for him.

Dr Amir had to believe it would go well. Baghdad, he reminded himself, had been harder than this. Back then, there hadn't been a friendly city like Erbil within 50 miles, with well-equipped hospitals and efficient bureaucrats.

This time he had a plan. He was checking in with the office every hour or so. They had every permission slip they could conceivably need. There were only a few animals to assess. After that, if it was possible, he would make plans for their treatment and extraction. It would be all right.

They drove through the outskirts of Erbil, passing into the grasslands of the Kurdish steppe. At every checkpoint they slowed down, waved at the bored soldiers with the Kurdish sun flag on their arms, and carried on.

The road turned like a slinky through the undulating landscape. Every so often, a village or string of shops cropped up alongside them. The houses were all small structures made of mud-brick or breeze blocks – tiny and decrepit compared to the cavernous half-finished skyscrapers of Erbil.

After an hour or so, a checkpoint loomed ahead. It was bigger than the others – covered with a large metal roof, four car lanes running underneath it. There were two soldiers manning each lane, and a few others sat on plastic chairs with their guns leaning against their sides.

Dr Suleyman stopped the car. He seemed nervous. Ahead of them was a stagnant queue of cars that ended at the checkpoint, a few hundred feet in front. 'There it is,' said Dr Suleyman. 'You've seen it. Let's go back.'

Dr Amir feigned a look of polite enthusiasm. 'Let's just go up to the checkpoint and see what it's like,' he said. 'We'll be really quick.'

Dr Suleyman was getting more uncomfortable by the second. 'You said we were just going to have a look,' he said.

Dr Amir grinned out of the window as they rolled up to the checkpoint. Suleyman took over, exchanging pleasantries with the soldiers in Kurdish. They spoke for a moment, then Dr Amir got out the car and walked over to the checkpoint commander.

'I sent ahead for permissions to cross into Mosul,' he said, showing the commander their papers. 'We'll be back this afternoon.'

Everything seemed to be in order. Beaming with his good fortune, Dr Amir sat down in the car again, a dispirited Dr Suleyman sitting next to him.

'Don't worry,' soothed Dr Amir. 'It'll just take a minute.' This was not true. Both of them knew it, but neither said anything more.

Then they were off. They had crossed the line that marked the border between Iraq proper and the semi-autonomous Kurdish region. As they drove, the land grew sparser, less obviously tended, and a burning smell stung their noses.

They kept to the far side of the road. Every ten minutes or so an armoured Humvee, painted dark beige or black, would screech down the road next to them, the few displaced people who had made it this far scattering as it approached.

At each checkpoint, they stopped and – after fielding lengthy enquiries about their health – explained what they were doing. Most of the soldiers, now with the Iraqi flag pinned to their tac vests and their upper arms, waved them on with good wishes for the animals. When they didn't, Dr Amir would produce a letter from Four Paws – stamped and written in English that few of the soldiers could read or speak – and drop a few names from the Iraqi authorities.

Another checkpoint, and they started to see them: thin streams of people, in groups of around ten, walking slowly towards an assembly point that seemed to have been decided on by the first person who sat down there. They were displaced people who had made it out of Mosul, and were fleeing the fighting.

There were more children than adults. Men carried old women on their backs as toddlers pootled alongside them. Only the adults looked scared, dust-streaked faces staring ahead. When they sat down, there was no relief, just a collapsing of rag and bone. The children, some screaming with laughter, chased each other around the side of the dirt road, ignoring the sweat and dirt on their too-hot velvet dresses and thick jumpers.

The women were wrapped in layers of black that covered their bodies and their hair. Most showed their faces, a black square of material over their forehead where they'd folded back their face covering. They were all oddly pale. For two and a half years, barely any of them had felt the sun on their faces.

The checkpoints were fewer now. There were barely any civilian cars on the road – these had been banned by the army, who feared the suicide cars – so they travelled alone in the ghostly quiet.

Soon the road was banked by broken houses; great implosions of crushed glass and twisted rebar that had been homes until the war came. Everything seemed dead apart from the streams of people – some still carrying white flags of surrender – hurrying past them.

They were approaching Mosul's eastern suburbs, where fighting had raged just a few weeks before. Unlike the outskirts, which had been flattened with airstrikes, here – in some areas – soldiers had fought house to house with militants, leaving the walls scarred with sniper bullets and the roads pockmarked with craters. Every so often, they would see a dead body by the side of the road, still in the dust.

As they drove, the doctor scoured the print-outs of satellite maps he carried. On them, the zoo's perimeter was clearly marked, as were the roads that led to it. But amid the chaos of liberation, buildings had been flattened, roads destroyed and neighbourhoods given complete architectural makeovers.

After half an hour of crawling through the streets of east, they reached the 15-foot chain link fence that surrounded the zoo. By the iron gates hung a sign: 'Welcome to Mosul Zoo'. It was decorated with pictures of a bear and a lion. Both of their faces had been spray-painted over black. Old plastic bottles and smashed-up masonry crunched under the tyres. They heard the not-so-distant sounds of gunfire.

As they slowed down, a squat, bearded man ran up to them. He was grinning, and looked genuinely pleased to see them. He didn't seem to notice the crashes and rifle shots that had left the vets flinching in their seats. Beaming, he waved them through the gates, and they parked.

It was, Dr Amir thought, as he clambered out of the car, quite a nice place. Through the trees, he could see a pond and a large gold statue of a coffee pot, but he couldn't see any animals.

'There used to be ostriches there,' shouted the man who was striding towards him. 'But they died of shock.'

The man looked about sixty. He had a shock of orange hair greying at the roots and a voice that carried like the cholera.

'He reminds me of my father,' Dr Amir later remembered thinking, as the man crushed his hand.

'Assalamu aleikum,' said the man.

'Wa alaikum assalam,' said Dr Amir, smiling.

'I'm Abu Laith,' said the man. 'Are you a doctor?'

'I am Dr Amir,' he replied. 'I'm an animal doctor.'

Abu Laith laughed. 'Let's go,' he said. 'I want you to meet Zombie.'

The man, tailed by Suleyman, Dr Amir and – at a respectable distance – a convoy of small children who'd appeared out of nowhere, marched through the gates into the zoo. The doctor, hair coagulating in the dust-heavy air, wished he had his instruments and his team with him. He hadn't brought any medical supplies, as he hadn't known whether the animals would be alive, how many they would be, and whether it would be safe enough to treat them. This trip was only to make an assessment of their condition.

The zoo was dotted with splintered and fallen trees covering the paths around the zoo. They saw the bones of a dead animal in the bottom of a cage by the entrance, and an unexploded grenade under a tree. Bullet casings were scattered on the path.

Abu Laith, still beaming, kept up a steady stream of information boomed at a level usually reserved for speaking at loud concerts. 'Another lion died this morning,' he said. 'She was the mother of Zombie. She had eaten Father, and he poisoned her, and she died too.'

Dr Amir's mind was 20 feet in front, where three small cages held a live lion, a dead lion and a bear. The reek of urine and old diarrhoea told him it had been weeks since anyone last went inside to clean them. They were far too small for the animals, he thought – probably bought for them when they were cubs. The bear was breathing in short, shallow whimpers as she lay in the weak sun. The living lion was panting. The bars above them were caked in strings of dried rubbish – food remnants lodged there as the children threw in the fruit and old meat. In another cage lay an older lion, emaciated and very clearly dead.

'I'm sorry, doctor,' said Abu Laith. 'We tried to help them. I raised Zombie with a bottle. He is like my son. And I tried to save Mother, but she was poisoned by Father when she ate him.'

The smell from the cages was unbearable. Amir's eyes were stinging. The deep zoom and crash of artillery sounded very close. Dr Suleyman was all but tugging at his sleeve.

'They need a lot of help,' he told Abu Laith. 'I'll be back in two days. Keep them alive until then.'

Abu Laith, cringing with shame, promised he would.

'We're going to have to take them out of the cages,' said Dr Amir almost exactly forty-eight hours later, as he prepared the blowpipe loaded with sedative he had brought with him from Austria. Hakam had come this time, and was wearing rubber gloves and a paper mask.

The doctor spun around, halting the crowd that followed him. 'You,' he cried. 'Grab the saline! You, where's the detergent? You, kids, you need to wash your hands if you're going to help, so stop gawping and get to work. We'll need more of you. Go and get the neighbours' kids too.'

The enclosure boiled with activity. Abu Laith all but saluted Dr Amir as he set to work. Within minutes, the children had formed a line to the well, chain-lifting buckets of water up to the cages.

Abu Laith held up a white sheet that was to be Zombie's stretcher. Dr Amir, who had been fiddling about in his briefcase, produced a blowpipe, took a stance like an assassin and soundlessly shot a dart at Zombie, then at Lula. A few shallow breaths, and their breathing became evenly soporific.

The doctor craned at them through the bars. It would, he knew, have been too risky to put them under completely. They were still in a war zone, and the animals looked close to death. Complete anaesthesia was out of the question. He could only sedate them for half an hour at the most. They would have to be fast.

'Open the cages,' he cried, and they snapped open Lula's cage. The bear looked small and harmless, Abu Laith thought. 'We only have half an hour, so you need to be quick. They need to be back in the cages before they wake up.'

With a snarl, Abu Laith scattered the crowd who had gathered round to take selfies with the lion and gawp at the doctor. 'Make yourselves useful,' he said. 'Stop staring and start cleaning.'

Pushing her hot fur with their hands and elbows, Abu Laith and Hakam – wearing paper masks, at the insistence of Dr Amir – rolled Lula on to the sheet. They dragged her into the courtyard.

Soon Zombie lay next to her. Dr Amir and the Four Paws team called him Simba, his intended name, though Abu Laith rigidly stuck by Zombie.

'Stand back,' shouted Dr Amir. Everyone listened apart from Abu Laith, who had immediately assumed the post of right-hand man to the veterinarian.

'These animals are in a very bad situation,' the doctor said. 'I am going to inspect them now. I need to get very close to touch them, to hear their heart and their lungs.'

Dr Amir bent down to examine Lula with practised fingers. 'Conjunctivitis,' he pronounced, pulling back the skin of her lower eyelid, the lashes sticky with pus. 'Severe dehydration.' He probed her abdomen. 'Malnutrition,' he said. 'Joint problems. Tooth problems.'

He looked at Abu Laith, who was standing next to him. 'How long has it been since they had the right kind of food?' he asked.

'A few months,' said Abu Laith, deeply embarrassed. 'But it wasn't because we didn't try. We couldn't go into the zoo for a long time.'

Dr Amir smiled. 'She'll make it,' he said quietly, still shaken by the thought of how much Abu Laith reminded him of his father, a pharmacist who still lived in Egypt.

Zombie, it transpired, had fleas, malnutrition and parasites.

From his case, the doctor drew three syringes. He filled the first one from a small bottle of antibiotics, and pushed it under the lion's skin. He took the second and the third, shooting them in the belly fat, hoping the anti-parasitic inside would kill the creatures that had nested in rings all over the worn ochre coat, and help Zombie recover from his neglect.

The animals' bellies and their feet were scabbed and weeping from lying in their own filth. Urine, after prolonged contact with flesh, works like salt and acid, eating its way into wounds and causing agonizing pain. Zombie had scratched his wounds, making them even worse. Working carefully, Dr Amir cleaned them, and the clumps of black, tic-infested hair inside the lion and bear's ears. He listened to Zombie's heart, which was damaged from the long-term infection. Hakam sat nearby, filling syringes according to the doctor's instructions and passing them over.

More than anything, the doctor could see, the animals were traumatized, flinching at every sound. 'We are going to try to save the animals,' he announced. 'We're going to treat them a little now, and then I am going to come back and take them away.'

He turned to Abu Laith and the children, who had crowded around him. 'This is an important job you have here,' he said. 'You are responsible for keeping these sick animals alive. If you do as I say, they'll live. If you don't, they'll die. We need to clean the cages with water and soap. You need to use brushes and scrub them very, very hard to get all the dirt off.'

The kids agreed.

Dr Amir turned to Abu Laith. 'Without you they would be dead,' he said. 'Now you have to keep them alive for a little longer. Every week, Zombie needs thirty-five to forty kilos of meat and as much clean water as he can drink. Don't just give him meat, give him bones, too, so he gets minerals like he would if he was hunting in the wild. Lula needs eight to ten kilos a day of fruit and vegetables – cucumbers, tomatoes, whatever you can find. She also needs honey.'

'On top of that,' continued the doctor, 'you'll give them these medicines once a day, crumbled into their food. If you don't do even one of these things, they could die.'

A while later the doctor, Hakam and Abu Laith sat together in the zoo, writing down dosages and going over their diet plans. Zombie and Lula were stirring in their piercingly clean cages, which reeked of detergent. Every day, they needed antibiotics, vitamins, anti-parasitics and anti-inflammatories crumbled into their food.

'Hydration will be the key,' said the doctor. 'If you can get them enough clean water, that is the start.'

Abu Laith lit up. 'And whisky, doctor?' he asked. 'I used to feed whisky to the dogs when I was in the army. If you give me one week I'll get Zombie used to whisky. It'll make him strong.' Though he didn't tell the doctor this, Abu Laith had, very secretly, already tried feeding Zombie whisky when he was a cub. Once, he had injected it into an apple and given it to him. He liked it, Abu Laith thought, but others might not understand. Before Isis came, they had whiled away long nights together in the zoo by the river, Abu Laith sitting outside Zombie's cage looking over the wide, slow churn of the Tigris, sipping from a bottle of Black Jack. The occasional slug went into Zombie's water tray, and it seemed to make him happy.

Dr Amir wasn't entirely sure what to make of this, and considered for a moment before discouraging Abu Laith, in the strongest terms, from ever giving whisky to anyone other than adult humans. Instead, he gave him some money to buy food for Lula and Zombie.

Hakam, it was decided, would be in charge of other acquisitions. New cages, spacious and shaded, would be built for their next big journey. The doctor put new locks on the existing cages, so that no one could come and steal the animals.

'You,' said the doctor to Abu Laith, 'will be responsible for their wellbeing. You'll protect them.'

'I already do,' said Abu Laith.

'I know,' said the doctor.

The sun had started to dapple the ground. Dr Amir was stretching his tired back and putting away his blowpipe. Then a thick blast made the ground beneath their feet shudder. The two doctors jumped. No one else did.

'Bomb car,' said Hakam, conversationally. It was, the doctor reflected, time to go. Abu Laith shook Dr Amir's hand. It pained the doctor to leave the animals like this. They couldn't understand what the bomb blasts meant. To live unmoored from the human processing of tragedy must, the doctor thought, be horrible.

He turned to Abu Laith. 'I'll be back as soon as I can,' he said. 'When we come, we'll be taking them away. You need to remember that.'

'I know, doctor,' said Abu Laith. 'They are wild animals.'

But the doctor knew, and so did the former mechanic, that things were never that easy.

44

Dr Amir

On a cold day in March 2017, Dr Amir's team packed their bags and flew to Erbil. Marlies, Dr Amir's assistant, was extremely nervous. She was an emphatically pleasant young woman, who had left behind a career as a mayor's assistant in Burgenland – a bucolic Austrian province near the Hungarian border – to join the madness of Dr Amir's life. Within a few months, she had gone from a relatively cloistered existence to one where she would turn up in Myanmar save sick water buffalo at a moment's notice. She had never been to a war zone. But when the doctor had asked her to come with him to Mosul, she had said yes, knowing that he needed her there. She had pored over the pictures of Zombie and Lula, horrified at their condition. But she was afraid that she would crack at the crucial moment. The only thing she could do, she surmised, was to carry on quietly and not say anything. The day that she had got the job, she had promised to trust the doctor completely. Now it was time to put that into action.

They left the office in Vienna weighed down with bags of assorted medical instruments, and landed around 8 p.m. The city was dark and the air was thick with a caking dryness.

It was a motley team that piled into the car at the airport along with Marlies and the doctor – who had flown back to Vienna briefly after his initial visits to the zoo. Yavor, a veteran employee of Four Paws Bulgaria, would handle communications with Vienna. Gregor, an Austrian cameraman, would be in charge of editing and producing media packages around the rescue. He had been told in no uncertain terms by his line manager that he had to stay in Erbil, and would be going nowhere near Mosul. He was not excited at the prospect of being so far from the action. Then there was Anton, a South African cameraman with extensive experience of Dr Amir's unusual ideas.

They had a team. But they still didn't know whether they would be allowed to enter Mosul and rescue the animals. When Dr Amir had gone there with Dr Suleyman, they had been two Arabic-speaking men in a small car. The frontlines were still fluid, and the borders chaotic. Now, the situation had stabilized. Getting a truck with Europeans and two predatory animals out of a city where the soldiers feared car bombs and foreign fighters was not going to be easy.

It was, however, a situation that was perfectly suited to Dr Amir's persuasive skills. First, after checking with Abu Laith, he had called Ibrahim – the owner of the zoo animals – who lived in Erbil. Ibrahim had owned the animals when they lived in the Forest, and still technically owned Lula and Zombie. Before Isis, he had entrusted the management of the zoo in the Forest to Ahmed, and their wellbeing to Abu Laith, who looked after Zombie and was planning on starting a zoo of his own next door. Now, Ibrahim knew that the zoo had been forced to move to a park in the east and that most of the animals had died, but not much more. What's more, Dr Amir thought, he didn't care.

Ibrahim had been irate when Dr Amir got in touch. He had, it transpired, been on the receiving end of a number of phone calls from concerned Mosul residents – including Abu Laith – complaining about the state of the zoo and the hungry animals inside it.

'Take them,' he had fumed to Dr Amir.

The vet had started to explain. 'We've paid to feed them and keep them alive,' he said. 'If we hadn't fed them they'd be dead.'

Over the course of their conversation, one thing had become clear: for Ibrahim, the animals were nothing more than a money-making enterprise. Lula and Zombie were once-lucrative objects that had lost their value.

Dr Amir tried, abortively, to update him on the animals' welfare, but the owner didn't seem to care. 'I want to take them off your hands,' Dr Amir had said, sensing that this would be the way to Ibrahim's heart. 'They'll be gone, and you'll never hear a thing about them again.'

Ibrahim had assessed the non-existent downsides. 'I don't care,' he had said, Dr Amir later remembered. 'You can take them. This is a war. There are other priorities. I have a friend who lost all of his family in one missile strike. He was a doctor, and his wife and all of his children were killed. Why are you talking to me about the animals?'

It was settled. That evening, they checked in to the Classy Hotel, a proud institution of the city's Christian quarter beloved by journalists, war tourists and – from a distance – European diplomatic staff, who were banned from visiting it. Their security advisors had pointed out that the enormous plate-glass windows facing directly on to a busy road would implode if a car bomb was detonated, killing all the drinkers in the permanently busy bar. Despite being aware of this, none of the Classy's patrons ever seemed to mind.

The group had been picked up from the airport by the receptionist at the Classy. Once they had unpacked the doctor, Marlies, Gregor and Yavor commandeered a large table in the bar and called a meeting. A succession of security consultants, friends of army commanders, fixers, acquaintances and nosy journalists called by their table through the evening as they worked their way through tiny cups of coffee and a constantly replenished bowl of salted nuts.

Rumours surfaced and died with every new arrival. The border between the Kurdish region and Iraq proper was closed, one journalist said. No one could cross. Someone else said it was open, and that they had gone through that morning. It horrified Marlies, who was tossed back and forth from elation to despair.

For Dr Amir, who had heard rumours from Sofia to Saigon, it was nothing more than the usual murmurings of war. They would, he thought, find a way.

As the smoke from a hundred cigarettes curled into the air, the team made lists – their phones and notebooks filling with water measurements, medicines and route maps.

By midnight, when the caffeine was wearing off, Dr Amir stood to make an announcement. 'Everybody,' said the doctor, and the table quietened. 'We're here now, and we've made the plan for tomorrow. But I want to make one thing very clear: if anyone feels uncomfortable, unsafe, please tell me. It's not a problem. You can stay here in the hotel while we go to Mosul.'

There was a second of silence. Then everyone – apart from a visiting blogger, who decided against it – murmured their consent. They were coming.

The next day, the doctor took a taxi to the headquarters of the security company he had hired to protect them on their way to and from Mosul. He and Dr Suleyman might have gone alone but now, with the expanded team, he felt a new weight of responsibility. Eastern Mosul was teeming with Isis sleeper cells, and the front lines were fluid. Many of the roads still hadn't been cleared of mines. Any object by the roadside might be an IED.

In the air-conditioned office, Dr Amir went through the plan in detail. Satellite maps showed where the front line was, and which areas had been cleared. Large swathes of Mosul, including the Old City and much of the western bank of the Tigris, were still under Isis control.

Every day, Isis sent drones armed with grenades or explosive devices over the river – releasing them on to army vehicles, tank turrets or groups of soldiers. They would, Dr Amir decided, need two armoured cars to get them in and out of the city safely, as well as a truck. At each step of the way, their coordinates would be sent through to the office in Erbil.

As he worked, his phone buzzed with messages from fixers, soldiers, the Ministry of the Environment and the Classy Hotel receptionist.

The security arrangements were just the beginning. Lula and Zombie would need new cages, replacing the filthy, half-broken shells which were all they knew. To hedge his bets, Dr Amir had left Hakam with money and instructions to have cages made in Mosul to some very exact specifications that would allow them to be taken into a cargo hold – essential if

Dr Amir's plan was to work. But he couldn't be sure that they would be finished in time.

With this in mind, the doctor had arranged to borrow a set of back-up cages from a private zoo in Erbil – a haven for the exotic acquisitions of Kurdistan's ruling classes once the trophy animals were too dangerous to pet. The cages weren't the right size for the cargo hold where he planned to transport Lula and Zombie, but they would do in a pinch. He arranged to have them delivered to the hotel, and rushed away to check on the truck that would be transporting the animals the next day.

Marlies had peeled off that morning to the pharmacy to fetch medicine they hoped would pull the animals back from the brink. Gregor had gone to a hardware store to buy wire-cutters. Dr Suleyman had been dispatched with Marlies to find vaccine books for the animals, in the event that they might be needed during the rescue attempt. Most importantly, he had been instructed to impress upon the Kurdish authorities that the animals would not be staying in Iraq and that they should be allowed into Kurdish areas with only the bare bones of transit papers.

As she walked through the pharmacy, Marlies couldn't help thinking that maybe it was all pointless. The lion and the bear had looked half-dead in the pictures. Maybe they would be gone by the time help arrived, or too ill to save. She tried to shake off such thoughts. The only thing she could do was to concentrate on the task at hand.

By the time the doctor was back in the Classy lobby, the preparations had become common, if muddled, knowledge among the hotel's residents. Dr Amir tried to fob them off. He didn't want a scene. Marlies sat at a table, zoning in and out of Arabic conversations she couldn't understand.

Dr Amir's phone buzzed relentlessly. His main contact with the army sent vague, reassuring updates. Others told him the road was closed, or offered to come with them, smoothing the path for a few hundred dollars.

'Everyone is making money from this war,' he told Marlies, as they sat at a table by the TV screens. 'Everyone is trying to sell you fake information. They just want money. Especially if you're a foreigner.'

At about 7 p.m., one of the doctor's contacts called him to say the cages were being delivered. He went outside just as dusk was falling and climbed in the back of the truck to inspect them.

A moment later, his head popped out of the back. 'Perhaps I should have mentioned this,' he said. 'But these cages need locks on them.'

45

Abu Laith

THE KILLING BEGAN AT NOON. IT WAS SUNNY IN THE LION enclosure and the air was thick with dust from the passing Humvees, making throats itch and noses burn. Abu Laith was standing, legs apart, over a despondent goat, beaming in a state of acute anticipation.

The man and the goat stood just inches away from the lion in its cage, who was looking with great interest at the horned creature in front of him. They were separated only by a set of bars.

Abu Laith had, Marwan later reflected, been infused with an industry and determination slightly frightening to the casual onlooker. The cleaning regime, which had ceased amid the fighting, was reinstated with a zeal that surpassed all reason. Marwan and the children had been kicked, shouted at and cursed for their lack of thoroughness. Initially mutinous, they had only to be reminded of Dr Amir's warning that the animals would die if their cages weren't cleaned. Now they were all but sparkling.

Lula was snoozing in the corner of her cage, her belly now rounded by better food and a twice-daily dose of honey. Abu

Laith had taken to pouring it straight into her mouth from a bucket.

For the first time in months, he had gone to the butcher with pocketfuls of dinars. The proprietor of the shop, used to the begging children, who he alternately beat away or fobbed off with scraps, had been impressed. In recent days, Abu Laith had become a regular customer at the illegal sheep market and haggled as hard as ever amid the dust and the blood, in the shadow of the ancient city walls. Today he had something unusual: a live goat he'd had to order in especially.

The wild. Abu Laith had obsessed over it for weeks now. Zombie was going to the wild. He would have to hunt in the very savannahs where Mother and Father had – presumably – been born, before being smuggled to Iraq through Egypt, like most of the zoo animals in Iraq.

Sitting in front of the National Geographic channel in the living room where they'd hidden from the bombs, Abu Laith soaked himself in the sights and sounds, the look of lions on the hunt. Dr Amir had told him how to feed and care for Zombie, but he hadn't left instructions on how to prepare him for his future.

Hunting, Abu Laith knew, was an art. It required stealth and bravery, but also the willingness to kill. He didn't know whether Zombie – who ate from his hand, and liked having his belly scratched – had it in him.

The goat, had it known all this, would have hoped Zombie didn't. Instead the creature stood there, trembling, about to be eaten. Abu Laith let go of its horns and stepped back. Delighted, he ran over to where Marwan was standing, looking on with extreme distaste.

Marwan then, and later on, wondered whether this was

really necessary. The old man had a gruesome way of training lions for the hunt, and one that – even to him – seemed remarkably unscientific.

Abu Laith's eyes were fixed on Zombie, who was staring at the goat. The lion took a step forward.

'Go on, Zombie,' breathed Abu Laith. He didn't know whether the lion had the killer instinct, but he wanted to believe it. There was nothing worse than a lion that had become a dog. A true zookeeper, he knew, encouraged the wild instincts of his charges.

With a silent lunge, the lion struck through the bars, grabbing the goat around its front haunches like a mousetrap on a terrible spring. For thirty long seconds, he smashed the goat against the bars with thumps that turned Marwan's stomach – deep, hard thwacks that made the goat scream. Then, to Abu Laith's cheers, the ruined creature folded in half. Zombie pulled it through the bars with a great snap, a sound like a raw carrot breaking.

'You are a killer,' hallooed Abu Laith at Zombie, who was tearing off strips of the goat's still-twitching skin.

Working expertly, Zombie lifted a section of the ribs and tore away at the thick, sinewy meat. 'Look,' shouted Abu Laith. 'Look at what he's doing. He's never had a goat before but he knows how to do it.'

Marwan, still feeling queasy, excused himself.

Abu Laith crouched down in front of the cage. 'You're a real lion now,' he said.

The lion would survive in the wild, he thought. But Zombie's departure still hung in the balance. A few days earlier, Ibrahim, the owner of the zoo where the animals had previously lived had called Abu Laith.

Someone had told Ibrahim that the strange Egyptian vet who had offered to take the animals off his hands was a rich man. The owner, Abu Laith later explained, smelled a rat. Why would anyone spend that much money to get the animals if they were worthless?

The more the owner thought about it, the more it must have seemed clear that the animals must be extremely valuable if this foreigner was going to make such an effort to get them out.

Ibrahim called Abu Laith, who he knew from before Isis, when Zombie had lived in the zoo by the river and the then-mechanic had raised him with a bottle as he waited for his own zoo to be built next door. Doubtless, Ibrahim thought he would be able to get the stocky ginger man on his side. What he hadn't accounted for was that Abu Laith was entirely devoted both to Dr Amir, who he saw as a guiding force in his life, and the animals.

'This doctor is a thief who is trying to steal the animals,' Ibrahim told Abu Laith, who was fuming on the other end of the line. 'Just make sure you stop him. I'll get the animals as much food and water as they need. We can buy another plot of land, and they can run free.'

Abu Laith considered hanging up. If they had relied on Ibrahim, Zombie and Lula would have been dead years ago.

As Ibrahim droned on, Abu Laith weighed his response. The easiest thing to do would be to tell him to get lost. But as he sat there, the seed of a thought bloomed. Congratulating himself on his cunning, he interrupted Ibrahim. 'The thieves,' he shouted, bringing the diatribe to a halt. 'I'm shocked. Don't worry. I'll stop them. I'll do anything I can. You can rely on me.'

And he put the phone down, rocking with laughter.

46

Dr Amir

The day of departure dawned damp and windy. Since the doctor's last visit to Mosul, the summer had turned to a harsh autumn, the rain washing rivers of waste water through Erbil's streets. As he crawled out of bed at 4 a.m., the doctor felt like he had barely slept. The night before, several well-intentioned people had told him in no uncertain terms that the road to Mosul was closed.

The border dividing Kurdistan from Iraq proper lay 20 miles from the outskirts of Mosul. Even if it was open, they would have to cross half a dozen army checkpoints before they got to the zoo. Each checkpoint was guarded by soldiers who had spent months at the front – making them jumpy and brutal by turns. New rules, or the rumour of them, could hold up a checkpoint for hours as it was escalated up the army chain of command or was ignored, leaving tired civilians and soldiers alike sniping at each other. For all their preparations, they were taking a chance.

By 6 a.m., Dr Amir was sitting in the back seat of one of the armoured cars after a brief breakfast of coffee, wrung out as a rag and racked with anxiety. In the front sat two security guards, dressed in the sweat-wicking khakis of their trade and heavily armed. Next to him sat Marlies, whose mind was strangely still as she stared at the bulletproof jacket in front of her on the floor. Back in Vienna, she had found the idea that they might have to use them both unlikely and bizarre. Now she lifted the jacket and found that it was extremely heavy.

Gregor and Yavor were in the other armoured car. The cameraman had been supposed to stay in Erbil and deal with communications, but decided against it. After many obstinate minutes of Dr Amir telling him that he would not be allowed to come, the vet had relented and taken him along.

The convoy contained another crucial element: Salah, a brooding member of the Iraqi special forces in his mid-twenties. He was Kurdish and had been hired by Dr Amir to introduce the group to the army and the Peshmerga, as he had contacts in both. The zoo was still on the edges of a war zone, and even with all their security guards and their pieces of paper Dr Amir knew that having the support of the army was the only thing that would matter.

Behind them was the truck with the empty spare cages for Lula and Zombie, now secured with padlocks and thick-linked chains. The new cages that Hakam had ordered for the lion and the bear fit specific size requirements that would allow them to be taken on the next stage of the journey. He had commissioned them from a welder in Mosul, who had assured him they would be ready by the time the doctor arrived. But in an abundance of caution, Dr Amir had bought spares rented from a private zoo in Erbil. They weren't the right size, but they would do.

The team set off, the air still light and fresh, cars clinking with medical supplies. By 8 a.m. they reached Khazir, a crossing point on the outermost edge of Kurdistan. There was already a queue snaking for half a mile in either direction, battered cars waiting sedately for the crossing to open.

'Keep to the side,' said Salah. They rolled past the waiting cars in the oncoming lane.

A group of Peshmerga soldiers were standing at the entrance to the covered area. Dr Amir's car slowed down as Salah's pulled over in front of the checkpoint. The soldier got out of the car, and Dr Amir followed him, papers in hand. A few minutes later, having explained themselves to a Peshmerga commander, they were back in the car, driving past the checkpoint. They were through the Kurdish territories, and the mood lightened considerably. Dr Amir checked in with the security consultants back in Erbil to tell them they had made it through the first checkpoint. The zoo was just 25 miles away. Dr Amir called Hakam, who sounded expectant, if slightly edgy.

The sun was starting to burn through the cold of the morning and, as they drove, it lit up the landscape around them in a gentle gold. Every mile their surroundings looked more decrepit and neglected. The area had for years been a no-man's-land on the edge of the so-called caliphate.

Now it had been taken over by the Iraqi army as they swept towards Mosul. When the army had come, Isis had fought in much of the village hinterland that lay outside the city, digging tunnels under abandoned houses, knocking through walls to flit along streets without fear of airstrikes. Bottles of their brown piss piled up near their sniper's nests as they waited for the enemy to come. All too often, the army had rolled through an empty village, only to find – days later – in a volley of bullets and shards of hot metal, that the fighters had been hiding below ground all along.

Grey villages passed them as they drove. They couldn't know, Marlies thought, whether these places were empty or not.

An hour later, they stopped outside a mansion that looked as if it was drawn from a dictator's handbook. It was surrounded by grey walls and in the dusty forecourt stood a brace of Humvees.

Dr Amir jumped out. They had come to this army base to procure an escort that would take them to the zoo and safely back again into Kurdish territory. It had been his idea, as he knew all too well how the checkpoints on the way to the zoo would relish holding them up. A gang of heavily armed soldiers would also, the doctor had reasoned, help to keep any obstructive locals out of the way when they were loading the animals on to the truck. He had seen too many times how a gathering crowd could turn violent. With two armoured cars and a truck, they were not exactly inconspicuous.

Marlies climbed out of the car, hair covered with a purple headscarf. The grounds stretched out in front of her, lush and green as if they'd been tended by a band of gardeners. How bizarre, Marlies later recalled thinking, that not long ago this had all belonged to Daesh.

A few soldiers crowded around, all talking to Salah. The doctor was greeting everyone with magnificent cheeriness. They were going to see a lieutenant general who commanded a large group of soldiers in this part of Mosul. One word from him could make Zombie and Lula's escape possible, or block their way.

Marlies caught up with Dr Amir outside the mansion. 'Anything we should do?' she asked.

'Everything is great,' beamed the doctor. 'Don't worry.' But privately, as he later admitted, the doctor was concerned. He couldn't be sure that the general would let them pass.

A soldier showed them through the front door and into a marble hallway with walls the colour of palest sand and a chandelier hanging from the high ceiling. They were ushered into a room with large sofas on each side that looked small in the cavernous space.

After a few minutes, the doors opened and a soldier walked in, carrying a gold coffee pot. He offered each of them a small, espresso-sized cup of very black coffee.

'Thank you,' said Marlies. She took a sip. It was some of the best coffee she'd ever tasted – rich in cardamom and somehow spicy and deep. This was, she reflected, a strange day. Everything seemed more alive than usual.

The doctor, who was used to these things, took his coffee outside and smoked cigarette after cigarette.

Half an hour later, a soldier came to fetch them. Marlies followed the doctor into the hallway. The soldier rapped on a huge pair of doors, opened them and led them in.

This room was even bigger than the previous one, ceilings soaring over a shining marble floor, high windows and enormous chandeliers. At the other end of the room sat a corpulent man with the familiar large moustache. He was wearing a beret and a large turquoise ring.

'Assalamu aleikum,' said Dr Amir, striding over to greet him with an air of bustling competence. Marlies and the rest of the group followed, grinning as hard as they could. The doctor had brought in the whole team with him, hoping to overwhelm the general with numbers – making it embarrassing for him to refuse their request.

They sat down, Dr Amir closest to the general, on a sofa inlaid with crystals. Beaming, he began to explain their mission. Better to affect an air of innocent exuberance, he later explained, when everyone would be expecting him to come with an ulterior motive.

'We're asking for some help for our mission,' he said. 'We have a big team.'

The general looked disgruntled when he heard about the plan to evacuate the animals. 'Why don't you take them to Baghdad instead of Erbil?'

'We would,' said the doctor, who would do nothing of the sort. 'But because Mosul is so dangerous for foreigners and since we're leaving so quickly it is just easier to take them through Erbil.' He had expected that this would come up. The Iraqi army answered only to the government in Baghdad, and were deeply suspicious of the Kurds.

Projecting a confidence he did not feel, the doctor kept pushing his case. A few minutes later, he fell silent. The Four Paws group looked at the general.

'Ahlan wa sahlan,' he said. Dr Amir grasped the general's hand. He turned to Marlies and the others.

'He said we're welcome,' he grinned. 'He said we can go.'

47

Abu Laith

Shivering in the cold, Abu Laith waited by the gates for the doctor to arrive. Already, everything was going wrong. A group of irate Moslawis had gathered outside the zoo.

After Ibrahim had spoken to Abu Laith, rumours had started spreading through the neighbourhood. A foreign thief was coming to steal the animals from the zoo and make his fortune by selling them on.

Word had soon got around – amplified in the chaos – that Zombie was about to be stolen and sold for a million dollars. It was impossible, the locals agreed, that such riches should be wasted on a thief who would sell their lion to another zoo at great profit. They had to stop them. Together, they debated where the thieves planned to take the lion. Some, Abu Laith later explained, said to America, where they would sell it to a New York zoo. Others thought it might be going to Baghdad, where the thief no doubt had connections to the government.

Amid them, sulking silently, stood two of the owner's men – gym-honed toughs who the onlookers thought had been sent to stop the thieves.

Abu Laith hovered around the edge of the group. He hadn't told Dr Amir about the rumours flying around the neighbourhood, fearing that it would make him reconsider the mission. Every few minutes, his phone rang with an update from the Four Paws team: a checkpoint passed, a new guide met.

Nerves, frayed to breaking point, were already starting to snap. Hakam, who had never met Ibrahim – who had lived in Erbil since even before Isis rule, investing his money in Mosul from a distance – was concerned, and decided it would be best to stay away from the zoo that morning, at least when Dr Amir first arrived. He had more pressing things to do, anyway. The cages he had ordered a month before would only just be ready on time, and he spent the morning cajoling the welder into finishing them.

Hakam planned to arrive at the zoo about half an hour before the cages were delivered there. He didn't want Ibrahim knowing how involved he'd been in the rescue process. Daesh might be gone, but it was still a dangerous time in Mosul. If he had the connections, Ibrahim could call in a favour and have him arrested. But Hakam had survived Daesh, and he was feeling pretty brave.

At the zoo, Abu Laith was nervous. He had been up since before the sunrise, preparing Lula and Zombie for the ride. More than anything, he was worried that Ibrahim's men would try to stop them being taken.

'You'll need to keep quiet in there,' he had told them. 'You'll need to lie down and sleep until you get to the wild. Then you'll hunt, and you'll run.'

Zombie's training had continued apace, and he could now

hunt chickens as well as goats. As the day of departure had grown closer, Abu Laith had started to worry whether he could really let Lula and Zombie go.

It was raining as Dr Amir's car pulled up at the front of his convoy. Two armoured cars full of soldiers followed them, stopping at the zoo and looking moodily out onto the park.

Abu Laith ran over to the group, and Dr Amir embraced him. They must look, the vet thought, like Laurel and Hardy, two jovial men – one tall, one short – gently cuffing each other in the rain.

'Are they ready for us?' asked the doctor. Abu Laith said yes. Amir noticed he looked extremely nervous.

None of the waiting locals had expected the army to come. Looking immensely bored, the soldiers climbed out of their armoured cars and strolled over to the gate. The hangers-on now became decidedly shifty. The owner's men pulled out their phones and started tapping frenetic messages.

Abu Laith showed Dr Amir through the gates. The others followed – the children scattering through the trees to the lion's cage. All of them were very excited to see the blowpipe with the sedatives again.

The entrance was blocked by the security guards who had come with Dr Amir – brawny men with guns. But that was nothing to the locals, who slid over the walls and through the smaller side door to watch the action. The two muscle-men followed. They were, Abu Laith assumed, updating Ibrahim on the rescue mission. One of them started filming the Four Paws team on his phone, Dr Amir later remembered.

Some in the crowd had been to the zoo during the Daesh occupation, and remembered the lions in their cages, and how noisy they had been at night. Others remembered how

Abu Laith had raised one of the lions from a cub, and how he would always look after a sick dog if you brought it to him.

Either way, the potential loss of the lion and the bear had begun to mean something. If they had been asked the previous day, the locals might have said that the animals should go to hell. Now it seemed unfair that they were being rescued, or stolen, or whatever was going on. It was just another way that choice was taken away from them, as it had been so many times before.

Dr Amir and Abu Laith – along with the doctor's team – were half-jogging to the cages, blowpipe loaded with a sedative dart at the ready. The fighting was close, the growing crowd was hostile and the vet knew they would have to be as quick as possible.

The children ran ahead of them to the cages. Abdulrahman was very sad, but he hoped no one would notice. In the rain, he fought against the desperate fact that Lula and Zombie were going, and he would never see them again.

His heart heavy, he sat on the ground next to Lula's cage. 'We're going to miss you,' he said. Hearing the others coming behind him, he stood up and ran into the trees, deeply embarrassed.

The rest of the group rounded the corner to the cages. Dr Amir knelt on the concrete floor and opened up his kit. He took out the blowpipe. Lula and Zombie were stirring in their cages, looking inquisitive.

Dr Amir was struck by the cleanliness and order of the predators' enclosures. The bear's belly was rounded, her eyes shining. Zombie's ribs were a little more cloaked in fat. Outside their cages, the concrete floor was still sparkling clean from the latest in what looked like a series of scourings.

'They look wonderful,' he said to Abu Laith. 'You did as I said.'

An explosion sounded nearby. No one reacted. Abu Laith, who had been standing by the cages, glowed at the doctor's words. 'Zombie can hunt,' he told Dr Amir. 'We've been giving him goats to kill.'

Faintly bemused, the doctor stood up.

The children were clustered in a corner by the monkey cage, seething with excitement.

Dr Amir took aim and shot a dart into Lula, who was standing with her paws up on the bars at the back of her cage. She shuddered, and was still. Pausing to reload, he shot another into Zombie's flank.

'Ready,' the doctor cried. Unlike last time, he had given the animals a dose that would put them under for at least half a day. But time was still short. It was the humans outside the zoo that worried him. All it would take would be for someone to push the security men too hard for the rescue mission to explode into violence.

Still more worrying were the drones that Daesh were still sending across the river – cameras attached – that hunted for groups of Iraqi soldiers.

Not five minutes after he had shot the darts, Lula and Zombie were both deeply asleep. Dr Amir unlocked their cages and went to the back of Zombie's enclosure. The lion's sides were heaving as he slept. 'Take this,' said Dr Amir, throwing the corner of one of the sheets to an onlooker. 'We need to get it underneath him.'

It didn't matter that he'd said goodbye a dozen times. As he watched the doctor drag Lula and Zombie out of the cages, Abu Laith wished that they were staying.

They gently placed the bear on the concrete floor, a few feet from Mother's grave, so that Dr Amir could examine her. His hands passed quickly across her fur, pulling open an eye, feeling her soft belly and running a finger along her ribs. He was doing the bare minimum to make sure she would be safe during the journey. They had to move fast.

As Abu Laith stood and watched, his phone rang. It was Ibrahim. 'They've come to steal the animals,' he shouted. 'You have to stop them.'

Abu Laith tried to adopt a concerned tone. 'I've been trying to stop them,' he said. 'But they've brought half the army with them. I can't do anything about it.'

Dr Amir, who was treating Lula and Zombie, turned around to listen. Abu Laith, catching his eye, winked theatrically.

'I'll try,' the zookeeper continued. 'But I can't promise anything.' He hung up the phone and grinned. 'Let's go, doctor,' he said, cheerily.

Dr Amir grabbed the lion's makeshift stretcher. Helped by the others, he hauled it through the zoo and the main gate into the road, where the army waited. Lula followed, also carried on a sheet. Next to them was the lorry where the cages that Hakam had ordered would go, as well as a crane to move them. But they weren't there yet.

'Where are the bloody cages?' roared Dr Amir.

'They'll be here any minute,' Abu Laith said, unconcerned.

Dr Amir looked at the old man. He was grinning contentedly at the animals. 'We need those cages,' said the doctor. 'We need to leave now.'

The crowd seemed to have followed them through the gate. Amid the shouts of the children and the sound of gunfire, one of Ibrahim's men was filming them again. Hakam, who had turned up not long before, assured the doctor the cages would be there at any moment. But Amir, looking over the gathering crowd, knew they had no time. They would have to put them in the spare cages and start driving, hoping the new cages would catch up.

Dr Amir ran to the armoured car, where the driver was waiting. They had to move quickly, he thought, or something else would go wrong. 'Get that crane moving,' he said. 'We need to get them in right now.'

A few minutes later, Lula and Zombie were inside the spare cages, breathing softly. Abu Laith squeezed his hand one last time into each animal's hot, oily fur as they were lifted up into the lorries by the crane. Abu Laith would follow them as far as Khazir checkpoint with the new cages, then – it was dawning on him – they would be gone.

The doctor was giving frenzied instructions to his helpers when Ibrahim's men sidled up. One of them was holding a phone, which he tried to hand to the doctor.

'I don't have time to talk,' Dr Amir snapped. 'We're busy.'

'Take it,' the man said.

Dr Amir considered the men for a moment. Six feet of muscle apiece, in tight black t-shirts and sculpted beards and tracksuits. They looked furious.

He took the phone.

'These are my animals,' fumed Ibrahim. 'You won't take them.'

The doctor, who had half an eye on the unconscious animals and the frenzy around him, tried to interrupt, but Ibrahim kept shouting. 'When you leave Mosul, we'll shoot you,' said the owner. 'People are waiting on the other side of the checkpoints. As soon as you leave, we'll kill you. You won't make it back alive from Mosul. You are dead.'

The doctor bit back his anger. 'I'll talk to you in Erbil,' he said, and gave the phone to one of the men, who tried to hand it back.

'You have to talk to Ibrahim,' he said.

Dr Amir didn't have time for this. He handed the phone to one of the soldiers instead. The doctor watched him as Ibrahim repeated his threats.

'We have our orders,' said the soldier. 'We're following them. I don't care who you know.' He hung up. 'Listen, doctor,' he said. 'We can protect you until the border with Kurdistan. But after that, you're on your own.'

For a moment, Dr Amir stood there. He didn't quite know how to take the threat. The owner could have him killed, he was sure of that. This was Iraq, and Ibrahim was a rich man. But he knew he couldn't lose focus. For the moment, the threat was in the zoo, not in Erbil. He had to stay calm.

Then he looked up into the sky, and saw something hovering right above him. 'Drone,' he shouted. 'Drone. Drone, evacuate, evacuate.'

Friend and foe alike ran panicking for shelter. Even if they didn't understand what the doctor was shouting in English, they could follow his arm gesticulating for them to run. The children scattered under the trees. Marlies ran to the armoured car, and the doctor ran after her.

The security guards had told Amir how it worked. First Isis would send the drones over the river – using them as spotters for artillery. Once they had found a target, they would pound it with shells. The people in the zoo had to get away, now.

The doctor looked up through the car window. There was nothing there. The screams had died down, and the zoo was silent but for the distant boom of the shelling.

Abu Laith could only stand still, trying not to bolt. He later on swore that Zombie had been crying, so he did too. He felt like a son was being taken from him.

The engines started up. An agony of reversing, and they were out of the gate, speeding up the road away from the zoo as the children ran after them, whooping and crying.

48

Abu Laith

Dr Amir's armoured car hit yet another snag in the tarmac, throwing all the passengers an inch or so into the air. Ignoring his fears, the doctor continued to stare ahead, scanning the road for soldiers, militants and stray dogs.

They had passed the last few checkpoints at a clip, barely slowing down as the soldiers raised their hands to them in greeting. Just ahead lay Khazir, the last checkpoint before Kurdistan.

They were only a few hundred feet from safety. But they needed the new cages if the next stage of the plan was to work. When they left the zoo, Abu Laith had said he would be five minutes behind them with the new cages. They decided to wait for him.

The convoy slowed down and stopped at the side of the road. Their military escort had left them on the outskirts of Mosul and now there were only three cars left.

Dr Amir called Abu Laith. 'Are you on your way?'

'Of course,' the zookeeper said. 'We'll be there soon. We've got the cages. I'm driving them up now.'

'Hurry,' said the doctor, who was feeling extremely anxious.

'Don't worry,' said Abu Laith, breezily. 'We'll be there soon.'

Half an hour later, Dr Amir called Abu Laith again. 'Where are you?'

'We'll be there in five minutes,' said Abu Laith, not for the first time.

'Is this an Arab five minutes or an actual five minutes?' shouted the doctor, whose stress levels were now unbearable.

'Five minutes,' said Abu Laith, and hung up.

Once they were with the Kurds, Dr Amir knew they would be safe. The Kurdish regional government was in a constant low-level war of bureaucratic attrition with the Baghdad-based federal government, which controlled the Iraqi army. The Kurds would like nothing more than to take credit for having helped rescue the animals. All the group needed to do was pass this checkpoint. But they needed to bring the new cages with them, or their plans for the animals' escape would be ruined.

They had been waiting at the road by the checkpoint for an hour when a small, decrepit lorry appeared in the distance. As it drew closer, the doctor could see the cages strapped on the back, and the hunched figure of Abu Laith sitting in the driving seat.

'We're here, doctor,' shouted Abu Laith cheerily, piling out of the lorry.

The doctor was already running around the back to direct the crane. 'Move,' he shouted at the rest of the Four Paws group, who scattered to direct the lifting of the cages, which needed to be moved over to their truck.

Abu Laith had just trundled off again towards Mosul when a car full of soldiers pulled up next to them. Dr Amir was directing the crane, and didn't see them coming. The new cages were secured onto the truck. The group would be ready in a couple of minutes, and over the checkpoint to safety in another five.

A man in uniform got out of the car and walked over to them. 'Assalamu aleikum,' he said, politely.

Dr Amir was so harried he barely responded. The new cages were on the lorries now. They were ready to leave.

'Let's go,' said Dr Amir, running up to the front of the convoy.

But the man in uniform was there before him. 'Stop,' he said. 'We've had orders from above. You have to take the animals back to Mosul.'

The doctor couldn't force his mind off the task at hand: getting his team and the animals through the last checkpoint. 'We have permission to enter Kurdistan,' he said, and dropped a few names.

The soldier's expression didn't change. 'We have our orders,' he said. 'All of your permissions are over-ridden. This comes from the top. It comes from the commander.'

Dr Amir felt sick. The soldier was – he knew – referring to the famously volatile commanding officer of the Iraqi forces in Nineveh province, where Mosul lay. The doctor had not asked him for permission to take the animals – instead securing a letter from the deputy governor of Nineveh and various army commanders. 'But we've got all the permissions we need,' he said. 'I can assure you.'

'Your permissions are cancelled,' the soldier said, curtly. 'We're taking you back to Mosul now. You're coming with us. We need you to dart the animals so they stay calm.'

It was too much for Dr Amir. He couldn't countenance the idea that they had come so far, only for the mission to be destroyed. The sun was setting, and he knew that there was only so long that the checkpoints would stay open. 'No we're not,' he said. 'The animals are coming with us to Erbil.'

The soldier bristled. 'Get their truck,' he said. 'We're taking the animals back to the zoo.'

Dr Amir tried to argue, but the snatch had already begun. As he jogged back towards the lorry, a soldier was directing the driver to turn around.

'Stop,' said the doctor. 'Get out and we'll talk.'

'Leave now,' shouted the commander, apoplectic. 'Leave now or I will arrest all of you.'

The Iraqi armed forces were not known for their sensitive handling of prisoners. Dr Amir considered their options. They were very limited.

He tried to look ingratiating. 'I'm sorry,' he said. 'We'll go now.'

With a last look at Lula and Zombie's truck, Dr Amir jumped back in the car and started the engine. Marlies sat terrified as they drove back up towards the Khazir checkpoint. 'Don't worry,' said Dr Amir, though he was very far from unworried himself. 'We'll come back for them.'

49

Dr Amir

By the time they arrived back at the Classy, the sun had long set. Leaving the car with the leather-jacketed valet, a Michael Jackson lookalike with mirrored aviators and fingerless gloves, they walked into the lobby through the defunct metal detector. On the wall was a picture of Masoud Barzani, the corpulent leader of the Kurdish regional government.

The bar, and the restaurant that spilled from it, was full of journalists and aid workers, as it always was. Some of them were dusty and extremely drunk. They were the writers and photographers. Others looked cleaner, with white teeth and immobile hair. They were the TV journalists.

Music tinkled from the speakers. Marlies and the others slumped down at a table. Dr Amir had been called to reception to answer a message. The rest of the team had barely spoken on the ride back. Occasionally, one of them would venture an idea, which the others would – politely – crush.

'Well, I'm getting a beer,' said Marlies, and Gregor and Yavor nodded their agreement. They could see the doctor, his back to them, talking to the receptionist, framed by the backlit faux-marble of the desk. As the others opened their bottles of beer, he walked back into the bar through the double glass doors. His face had lost its healthy brown glow. It was grey as plaster and the skin around his eyes was dark.

'What is it?' asked Marlies.

'There was a message for me at reception,' the doctor said, sitting down at the table. 'People had been coming to the hotel today to ask about me. Then someone rang. They said I should go and meet them at eight tonight. They gave an address. They said I should come alone.'

The team gaped at him.

'It's Ibrahim,' Dr Amir said. 'There's no other option.'

It seemed absurd that someone would threaten him for trying to rescue sick animals.

Marlies was adamant. 'Of course you can't go,' she said, matter-of-factly. Gregor told him he would be stupid to fall for it. Yavor agreed.

'I'm going,' said Dr Amir. 'I have to. We need to win him over.'

'How do you know it's him?' Marlies asked.

'It's him,' said Dr Amir. Ibrahim, he knew, was an influential man. His family owned a successful restaurant in the Christian quarter. They were close to the governor of Mosul. The kind of family, in truth, that could have him eliminated if they chose.

He couldn't avoid them for long. If he didn't go to see Ibrahim, they would find him anyway. And besides, no one, he told Marlies, leaves an address with a hotel receptionist if they're planning a murder.

The security company, when he called them, agreed with Dr Amir. This was not the mark of an assassination.

'Fine,' Marlies said, deeply unsatisfied with this turn of events. 'I'm coming with you.'

'No,' said Dr Amir. 'I have to face this. It's the only way.'

By 7.30 p.m. the doctor was standing outside the hotel flagging down a taxi in the dark. The others had, after the doctor explained himself, reluctantly agreed that he should go. The team would wait down the road in their own car.

The taxi wove through the back streets of Ankawa, the Christian quarter, the roads lit only by signs for beer shops. The doctor didn't feel truly scared, he recalled later, but he was very tired. He didn't know what would happen when he got to the meeting, but the situation was desperate, and this was the only way it could change.

He paid the driver and got out. They were on the side of a busy road, the headlights of passing cars glinting off the shopfronts around them. Dr Amir walked up to the restaurant named in Ibrahim's message. It was a huge, brightly lit hall of the kind very popular among Erbil's upper middle class. But though it was dinner time, it was completely empty.

He walked through the doors. A waiter, dressed in a white shirt and black waistcoat, came up to him smartly.

'Just one,' said Dr Amir, and the waiter showed him to a table. It was enormous, like all the others, and could have seated at least ten people. Dr Amir looked around. Thick gilt-glass chandeliers hung low from the ceiling. On one side of the restaurant, half a dozen waiters stood around, not doing much at all.

'Would you like to drink something?' said the waiter who showed him to his table.

'Thank you,' said the doctor. 'I'm waiting for someone.' He sat, trying to look calm. There was a camera on the ceiling, pointing towards him. Somebody, he assumed, must be watching from another room. He wondered when they would come.

Affecting a nonchalance he did not feel, Dr Amir lit a cigarette and asked the waiter for tea.

He was just taking a drag when a side door into the restaurant opened and a tall man with a bald head walked in. He was gym-honed and heavy set, dressed in a tight blue shirt and jeans – in his forties but looking well for it. Dr Amir knew, without a moment of doubt, that he was very dangerous, and very angry.

'Good evening,' said the man, walking up to the doctor's table. 'So you are this Amir Khalil.'

The doctor tried to look unconcerned. 'Yes,' he said.

The man pulled out a chair, and sat down at the table. 'I want to be very clear from the beginning,' the man said. 'You are in my country. If I say you stay, you stay. You can't go home without my permission.'

Dr Amir tried to appear calm. It was best, he thought, to stay quiet.

'Who are you working with?' the man snarled, leaning forward and smacking his hands on the table.

The doctor found his voice. 'Four Paws,' he said. 'It's an animal charity in Austria.'

'How do I know that?' interrupted the man. 'Where's your ID? You're not in Austria now.'

Dr Amir didn't know what to do. A thuggish, angry man was asking for his staff identity card, which he didn't have on him.

'I don't have it,' Dr Amir said. 'I've left all my cards at the hotel, and my passport.'

The man's irritation was almost palpable.

'But I've got my Austrian driving licence if you want?'

'Show me that you are Amir Khalil,' barked the man, stretching out his hand.

The doctor took out his driving licence and handed it over. The man looked at it for a moment, then put it in his own pocket.

It was a power play, Dr Amir thought, and tried to stay calm. 'Look,' he said, 'you're asking me questions, but we're not in a police office, or a security office. We're in an empty restaurant. Who are you?'

'You're trying to steal the animals,' shouted the man. 'Tell me the truth.'

The waiters stayed where they were, still and silent.

'I'm not trying to steal them,' said Dr Amir. 'I spoke to the zoo's owner and he said I could take them. They were almost dead and I wanted to rescue them. I'm not going to make any money from them.'

'Liar,' the man shouted. 'You're working with American intelligence. You're CIA.'

Dr Amir shrugged. 'As you can see, I'm Egyptian, not American,' he said, affecting a casual tone he did not feel. 'I'm just a vet.'

Accusations of working for the CIA were always the first, he knew. Soon they would accuse him of being an Israeli agent. That, he knew from experience, was when you knew things were really bad.

This couldn't be Ibrahim, Dr Amir thought. He was too young. He had to be one of his thugs, sent out to test him, or just to scare him.

'At 8.22 this morning you spoke to a man in the army,' said the man. 'We know all about it. What did you talk about?'

Dr Amir didn't answer. The man had been a contact of his who had helped arrange the security escort to the zoo. He hadn't told anyone else he'd called him.

The man looked triumphant. 'How do you know this guy?' he asked, leaning forward. 'Who gave you the contact?'

Dr Amir felt, for the first time, a touch of real fear. They had a record of his phone calls. This, the doctor knew, was out of Ibrahim's league. He was a rich but ultimately unimportant man. This was something different; only someone really connected would have access to this information.

'Who are Four Paws?' asked the man, shouting now, not bothering to wait for an answer. 'How did you get all this money? Who is with you? How much money do you have?'

Dr Amir had no idea who this man thought he was. 'I haven't got any money,' he said. 'I've only got my credit cards.' He tried to sound as steady as possible. The angrier the man became, he thought, the calmer he should be – unaffected by the accusations. If he became angry, it would escalate. He would look guilty.

'You're coming to spy on our country,' the man shouted. 'We're not going to let you leave. You've been working with terrorists. I know you've been in Gaza.'

Dr Amir tried to bring the conversation back to reality. 'You can just google me,' he said. 'That will tell you everything you need to know.'

The man ignored him. He had done this before, Dr Amir thought. The relentless questioning seemed to be part of an intimidation technique – something that he had, without a doubt, learned in some branch of the security services. This wasn't a civilian.

'Do you have children?' the man said, glaring at the doctor over the table. 'How many? Where are they?'

'I have three girls,' said the doctor, remembering vaguely that it was a good idea to try to humanize himself in situations like this. 'They're in Austria. They're Egyptian, like me.'

An hour and a half later, the man had yelled himself hoarse. Dr Amir had all but stopped talking. Everything he said seemed to infuriate his interrogator more. Silence was the one thing that calmed him. In the brief moment of quiet, the doctor lit a cigarette, inhaling a warm stream of nicotine that calmed him a little.

The side door opened again, and another man walked into the restaurant. He was in his sixties, straight-backed and proud, with a large belly and cotton-white hair. He sat down at the table next to Dr Amir. This, the doctor knew, must be someone very important. He looked like a sheikh.

'How are you?' asked the man, and his round, wide face looked not entirely unfriendly. Across from the doctor, the man who had been interrogating him was still seething.

'I'm OK, *alhamdulillah*,' said the doctor, praise be to God. The cigarette had made him feel significantly better.

'Doctor,' said the man, patiently. 'I can see you've made a very grave mistake. My brother owns the animals. He has told me what you tried to do with them.'

Something inside the doctor, who had been working so hard to remain calm, snapped. He started speaking, and couldn't stop himself. 'Maybe I did,' the doctor said. 'But the real mistake was made by your brother, not by me. I suppose he hasn't told you that I was in touch with him? That I'd told him what I was doing? That he'd said "screw the animals, take them, I don't care, I have other important things to do"?'

The owner's brother was sitting very still, staring at him, but he didn't interrupt.

Dr Amir' continued. 'I don't think it was a mistake to feed your animals,' he said. 'Or to give some money to the people who were looking after them, because you hadn't paid them. Is this really how you run your business? That I feed your animals, that I give money to the people who were looking after them because your brother didn't do it, and at the end I'm the one who has made the mistake?'

Dr Amir took out his phone and brought up pictures of the animals as they had been when when he first arrived – starved, half-dead and swarming with parasites. 'Look,' he said, showing the photos to Ibrahim's brother, who still had not introduced himself formally. 'Look how they were. They were almost gone. But we saved them.'

The man just smiled. 'Where are you from?' he asked, softly.

Dr Amir paused. The owner's brother was clearly the head of the family, and an important man. After setting his thug on the doctor, he wanted to play the benevolent Arab patriarch – reluctant to do business straight away, brimming with hospitality. Let him, then, play the game, the doctor thought. He would try to get him on his side. 'I'm from Egypt.'

The man smiled. 'Egypt,' he said. 'I like Egyptian girls. I've thought about marrying one.'

Dr Amir tried to smile as the man talked about Egyptian women. He barely heard what he was saying.

'Did you eat something, brother?' asked the man, when he had finished.

The doctor felt like strangling him. 'No, I haven't eaten,' he said. 'I thought there wasn't much point, since your man here kept threatening me. Why should I eat? If I'm going to die and go to heaven, I might as well be light, so I rise quickly. If I went after I'd eaten I'd be heavy, and it might be difficult.'

For the first time, the man cracked a real smile. 'We're not going to kill you,' he said.

The side door opened again, and another man came out. Smaller than his brother, slightly dilapidated in his baggy suit, a man who Dr Amir immediately thought must be Ibrahim cut a forlorn figure as he shuffled over to the table. Without looking at Dr Amir, he sat down next to his brother, who motioned for the doctor to speak.

'Isn't it true that I called you?' asked Dr Amir, looking at Ibrahim. 'I called you from Austria, and you told me that you didn't want these animals, and that they just gave you problems?'

Ibrahim was staring at the table, under his brother's gaze. 'But you took the animals without the permits from the commander,' he said.

'But that was the deal,' Dr Amir burst out, exasperated. 'I was meant to take the animals. I treated them.'

'It wasn't like that,' Ibrahim said, doubtfully.

The argument was forestalled with the arrival of a cart of mezze – hummus, salads and tabbouleh, pushed by one of the waistcoated staff.

'I'll eat over there,' said Dr Amir, knowing it would be rude to refuse the food. He wanted to leave Ibrahim to be told off by his brother. Played out, he sat down at a nearby table and started on the hummus. The brothers were talking quietly to each other. As he ate, the man from the security services stood up and came to sit down at his table.

After his earlier outburst, he seemed to have calmed down.

Dr Amir, sensing the time had come to befriend him, took the man's head in his hands and kissed his forehead, chuckling. 'Whatever else you have done, you are a good person,' he laughed.

The man smiled, and leaned back in his chair. He was staring Dr Amir deep in the eyes, and looked genuinely confused. 'Why do you want the animals?' he asked.

Dr Amir had seen it in Baghdad, in Cairo and in Bucharest. Some people just didn't care about animals, and couldn't understand why anyone else would. 'We want to take them to a better life,' he said.

The man smiled. 'Where are you taking them?' he asked.

Despite himself, the doctor was starting to get angry again. 'Look,' he said, firmly. 'Your friend told us he didn't give a shit about the animals. I went to go see them a month ago and they were almost dead. I just want to help them. It's what I do.'

The man looked appraisingly at Dr Amir, then leaned back in his chair. 'OK,' he said. 'You want to make a deal, let's make a deal.'

Dr Amir's heart lifted. The man looked at him, triumphant. 'One million dollars.'

Dr Amir felt faint. He searched for something to say. 'Just one million?' he asked eventually, in a high tone of disbelief. 'Is that what you really think lions cost?'

It was hard to know whether to laugh. The man looked very pleased with himself, as if he had outwitted Dr Amir in a tricky business deal. 'I'm not that easily fooled,' he said, smiling.

Dr Amir didn't know where to start. It was all so clear. The threats, the bullying, it had all come down to this crude financial imperative. 'They're not worth a million dollars,' he said, finally. 'They're not even worth a tenth of that. How much money did your friends buy them for? I'll pay that back.'

The man just smiled and shook his head.

Dr Amir was reeling in the face of such avid stupidity. 'If they go back to that zoo they'll end up dead,' he said. 'And then they'll be worth nothing at all. So if you're thinking of selling them, there's no chance.'

The brothers, who had stopped talking, came over to the table.

'Where are you going to sell them?' the man asked again. 'To America? To Baghdad? To Austria?'

'We're not selling them,' Dr Amir interrupted. 'All we're trying to do is save them, and take them to another place where they won't be in danger.'

The owner's brother stared at him, clearly amused. 'These are our animals,' he said. 'We invested in them. We lost a lot of money.'

The doctor, despite his baffled fury, knew what the man was doing. He was settling in for what he clearly expected would be a long negotiation over a fortuitous business deal. There was no point getting angry. The vet would have to talk his way out of it.

'Sir,' said Dr Amir, trying to smile as he addressed the older brother. 'Let me tell you a story.'

It was 11 p.m. by the time the doctor returned to the hotel. He had been gone for three hours. Marlies had waited up for him, while Gregor and Yavor sank bottle after bottle of cold Heineken. The table was scattered with the leftovers from their dinner. Her initial fears that the doctor would be abducted and killed had been assuaged by a message an hour after he arrived saying that all was well. But she didn't know what could be taking him so long.

Almost everyone else had left the restaurant by the time the doctor stepped through the door, looking immeasurably tired. He slumped down at their table. In the aquarium-tinted light, his face was drawn. One of the security guards, who knew what to do, handed the doctor a small bottle of whisky.

Dr Amir took a swig. 'It was the owner,' he said, finally. 'He's after money.' He told them what had happened.

'A million dollars?' Marlies half-screamed, as he spoke. 'But those animals are almost dead.'

'We went back and forth for so long. I told them they'd be heroes if they let us take them,' said Dr Amir, who felt like a broken man. 'But I don't think they care. In the end I just said we'd meet again tomorrow at 10 a.m.'

'Do you think they'll back down?' asked Marlies.

Dr Amir sighed. 'They'll start negotiating,' he said. 'They think we're trading at a market.'

'But can't you just explain that we're not selling them on?' asked Marlies. 'If they knew that we weren't making money from them, wouldn't that make a difference?'

'Of course I told them that,' the doctor said. 'But they just think I'm lying. They think that they trapped me while I was on the way to making the business deal of my life with their goods, and they want a cut of it. They've got a scent of money, and they feel I owe them. They'll keep negotiating until they're paid.'

Thoughts spinning, Marlies sat back in her chair, feeling utterly desperate.

50

Dr Amir

WINTER SHIFTED SLOWLY AS EVER IN NORTHERN IRAQ, THE wet disappointment becoming harsher, the winds biting harder. The flat land, beaten into cracked dust by the summer, had little time to recover before it was stifled with the thick ooze of Mesopotamian mud and river water, which had fertilized and flooded this land since the beginning.

The rain spattered on the roof of the portacabin by the Khazir checkpoint where Dr Amir and Marlies had been sleeping. The doctor sneezed uncontrollably. Everything reeked of dog shit.

Things were, all told, not going well.

Six days had passed since the soldiers had taken the animals back to Mosul, leaving the cages outside the zoo with Abu Laith, who was very happy to see them again. The next morning, against all predictions, Ibrahim had showed up only an hour late following the meeting in the restaurant. He had quickly dropped his asking price to a very small fraction of a million dollars. Dr Amir had agreed, with the stipulation that Ibrahim would sign a notarized agreement promising never to open a zoo again. Within the day, the sales contract had been signed, hands shaken and goodwill professed. Ibrahim called off his men in Mosul, and told them that the animals should be allowed to leave to Erbil.

But it wasn't enough. By denouncing Dr Amir to the army, Ibrahim's family had irreparably jeopardized the mission. Dr Amir's name had been put on all sorts of unpleasant lists, and the army were now dead set against the animals coming to Erbil. The commander's intractable 'no' still stood.

Undeterred, the doctor had taken Ibrahim to Mosul with him to attempt another rescue. The owner had sulked along in the back seat of the car, refusing to engage with Dr Amir's cheerful chatter, though the vet thought he seemed impressed by the scale of the Four Paws security operation. He had been insistent that he should go and speak with the army alone, and that he had contacts who would clear up the misunderstandings, and lift the block on the animals going to Erbil. From the Khazir Checkpoint, Ibrahim had gone ahead to see the army, who had sent him to the governor, who had sent him back to the checkpoint. He had gone home alone, still grumpy.

'The only person who can get you out of here is the commander,' the guards at Khazir had warned, not for the first time.

After extensive enquiries, Dr Amir discovered that the Khazir checkpoint was the only one on the road from Mosul that had been told not to let them pass. Without Ibrahim's men meddling, they could operate inside Mosul relatively freely. In an attempt to heap pressure on the authorities, Dr Amir went into Mosul again with a fleet of trucks and picked Lula and Zombie up from the street outside the zoo, where the army had left them. He brought them to Khazir, and parked the truck in front of the checkpoint.

That was six days ago. Now, just like the previous mornings, he and Marlies awoke cold and miserable, shivering in the thin wind. Lula and Zombie, sitting in their cages, were in far better fettle than they had been at the zoo – ploughing through honey, fish and haunches of beef each day, respectively, in their cages.

The doctor's clothes smelled like cigarettes from the men he sat with, every day, cajoling and befriending. When the guard changed on the evening of the first day, he had gone to drink tea with the new officer in charge, who had told him that he was sorry for his plight, but that there was nothing he could do. If the doctor had been Iraqi, he said, he would have been thrown in a military prison.

'You know my brother was killed here?' one of the soldiers had said. 'You know they killed him and cut his head off? Why are you here?'

Dr Amir had told him he was sorry, and that rescuing the animals didn't detract from their work as soldiers, or his brother's death. Others were more sympathetic, finding the group merely crazy, rather than offensive.

As it was, Marlies and Dr Amir were grudgingly permitted to sleep in an abandoned portacabin near the checkpoint. It was peppered with dog shit and reeked with a dank smell that made the back of their throats itch. The crimson fleece blankets inside were damp and so smelly that neither the doctor nor Marlies could use them. Yet as they sat in the cabin that first night, rain spattering on the roof, they were overcome by hysterical fits of giggling.

Marlies was tired, but kept her spirits up. She had never been to Iraq before, but had every confidence that the doctor knew what he was doing.

Dr Amir was feeling much worse. He had – in fact – very little idea of what he should do, though he was doing a passable job of pretending otherwise. He was sick with a horrible cold and incredibly stressed, fielding hundreds of calls to politicians, police, soldiers and the media to try to push them to open the checkpoint. With every hour that passed, he felt that their chances were slipping further and further away.

Dr Amir had sent the rest of the group back to Erbil to rest while he stayed at the checkpoint with the animals. Marlies, who saw how stressed he was, had insisted on staying with him. On the first night, they had fallen into an exhausted half-doze on the floor of the cabin, rain pounding on the roof. The doctor had a cough, and a fever that was growing higher.

Not to be discouraged, he had spent the next day calling everyone he could think of and visiting the governor of Nineveh, a large bald man famous for his furious outbursts at civilians asking for help.

Everyone had expressed great sorrow at Dr Amir's predicament, and assured him that everything would be fine. No one, yet, had explained how that would happen. At every juncture, he was given a new list of things he needed to provide in order to get permission to cross the border. One official told him that every living thing crossing into Kurdistan needed security clearance.

'You need me to prove that the lion and the bear don't

belong to Isis?' the doctor had shouted in exasperation. As he remonstrated, Zombie and Lula grew fatter – if faintly confused – on the meat and frozen fish that the soldiers bought for them with Dr Amir's money from the market.

When he begged to speak to the commander, or for others to approach him on his behalf, he was met with elaborately polite refusals. The commander was the famously irascible head of the entire Nineveh military operation, and no one wanted to anger him.

'Soon, God willing,' the Mosul governor smiled, when Dr Amir asked when they would be able to cross. The message was clear: they were stuck. Disobeying the commander was tantamount to death.

By the time the sixth day dawned, they had made no progress. All roads, it seemed, led to the commander and his intractable no. The doctor's requests for meetings had all been ignored. Messages left were never returned.

After days sitting in the cabin, rancid blankets huddled around their shoulders, the doctor's phone rang. It was a contact in the army whom he had begged for help. 'The commander will see you,' he said. 'But, please be careful. Between us: he's told his generals that if these animals leave Mosul, he'll put *them* in the cages with the lion and the bear.'

'Just tell me where and when,' the doctor said.

After lunch, Dr Amir washed his face and put on a new shirt. Jumping into one of the empty lorries that had been part of his convoy, he drove back through the checkpoints far into Iraq proper, waving merrily at the soldiers as he passed. He had brought Gregor with him so that he would have a witness, if nothing else.

Following the directions the doctor had been given, they came to a blast wall guarded by men dressed in the black uniforms of the special forces. Smiling, the doctor introduced himself and, after frisking them, the soldiers opened the gates and scuttled them inside.

It was an important military base, Dr Amir could see, protected with tight layers of security, checkpoints and razor wire. There were Americans here, too. He could see their helicopters.

A soldier sent them to a large building with big windows. Dr Amir and Gregor walked in through a hall to a meeting room, where about a dozen impressive-looking men were sitting at a conference table.

They sat down, and one of the soldiers brought them tea. It was all very civilized, but the commander wasn't there, and no one was talking.

A soldier came into the room. 'He's coming,' he said.

The men in the meeting room were kicked into action, and they piled out of the room into the courtyard, the doctor and Gregor following.

The doctor looked into the sky, and saw two helicopters flying towards them. One was American, he thought, and the other Iraqi. He ducked down as they came in to land, blowing dust over the yard. The soldiers surged towards the first helicopter.

As the blades settled, a man stepped out in a khaki uniform and a burgundy beret. He was tall, with a thick moustache – hard, stony and unbreakable.

'Assalamu aleikum,' said the doctor, pointlessly quiet amid the helicopter's clacking. He walked forward, shoulders thrown, holding out his hand.

The commander strode straight on past him, surrounded by his guards – big men moving sinuously as cats. An air of well-trained dynamism surrounded them.

Dr Amir stepped back. He thought it best not to say anything. The wind from the helicopter was fading, and the air was quiet. He turned and followed in the footsteps of the commander, his confidence deflated. By breaking into a faint jog, he had almost caught up with them by the time they reached the barracks. He had no idea how he was going to buttonhole the commander.

The decision was made for him when one of the soldiers turned around, blocking the doctor's path.

'Wait,' he said, pointing at the meeting room where they had sat before. 'Go and sit in there.'

The commander had already disappeared into another office. Dr Amir considered pressing his case and thought better of it. He walked over to sit at the conference table. There was little point, he reflected, in making them any angrier. But he would have to think of a plan.

After half an hour – during which Dr Amir smoked and watched as some soldiers drank tea – four generals filed into the meeting room. The commander was not among them.

'Dr Khalil?' barked one. The doctor stood up to greet him, only to be fixed with a furious stare. 'You're here again,' the general spat. 'With your lion problem.'

Dr Amir tried to look calm. He knew that the only way he would be able to get through this was if he stayed icily polite. They saw him as an Egyptian vet. He needed to remind them that he was part of a big, powerful European organization that was determined to get the animals out, one way or another.

'I don't have a lion problem,' the doctor said. 'You are the ones creating the problem. I respectfully ask that you let them pass into the Kurdish area. What am I supposed to do? Should I just let them die? Would this make you win the war, if they did? No, it would change nothing.'

The general interrupted. 'We've had to stop the war for two whole days because of your animals. Do you even know what you've done? Do you know what we're doing here? Daesh killed women, they killed kids. You understand? And you're coming here talking about your fucking lion and your bear? Do you think we have time for this shit?'

The man was warming to his subject. 'We are fighting a war,' he said. 'Our men are dying. We are taking Mosul back from Daesh. And every day people call the commander, asking him to stop the war for your stolen animals?'

'But I'm not asking that,' said Dr Amir, who realized, by the general's apoplectic expression, that he had spoken too soon. 'I'm just asking you for one kindness. I'm just asking that you open the checkpoint for me to leave with the animals. They'll be gone and you'll never hear from us again. We didn't stop the war for two days, and if we did, then please let me know how, and I'll try to stop the war for ever.'

He had barely finished his sentence when the general cut in: 'You're a thief,' he screamed. 'A thief and a liar.'

'Sir, you have the wrong information,' the doctor said. 'The animals belong to us. We're the owners. We have a sales contract.'

The general looked ready to burst. 'We know you stole the animals.'

Dr Amir fought hard to keep his voice steady, trying to look the general in the eye. 'I just want to rescue two animals,' said Dr Amir, speaking the most formal Arabic he could muster, making the generals sound like villagers in comparison. 'Just to take a bear and a lion away with the zoo's permission, and never come back.'

Another man came in, looking – if possible – angrier than the others. By the insignia on his uniform, Dr Amir could tell that he was the commander's deputy.

He loomed over the doctor. 'You know how to rescue these animals?' said the deputy. 'I'll tell you. There's only one way to rescue them.' And from his pocket he drew out two bullets and dropped them into the doctor's hand. 'Here,' he said. 'Take them and kill your animals, solve your problem.'

The doctor took the bullets, and held them as the deputy stomped back out of the room. 'I do not solve my problems by killing,' the doctor said loudly, as the door slammed.

The meeting was over. Together, the remaining generals ushered them back into the courtyard, to where their car stood waiting.

'That did not go well,' considered the doctor, as he started the engine. 'We'll have to try harder.'

51

Dr Amir

AROUND LUNCHTIME ON THE EIGHTH DAY, DR AMIR walked up to the guards at the checkpoint. The winter sun shone weakly on the muddy ground, and the steam from the tea-kettle drifted into the sky. Everyone smelled extremely bad.

'I've had enough,' said Dr Amir to the guards at the checkpoint, as they sat around with their cigarettes. 'We're leaving today.'

The guards all but gasped. 'Are you sure?' one of them asked. 'You're going to take the animals back?'

Dr Amir nodded gravely. 'Yes,' he said. 'We have no other choice. I might try again later when we have our permissions.'

The men agreed. 'It's much better if you have permissions,' they said, wisely. 'It'll be much easier.'

Head hanging, Dr Amir went back to the cabin where Marlies was waiting, sniggering like a naughty child. 'Nice work,' she said.

Dr Amir was pleased. He had a new plan.

The last eight days had been about as enjoyable as drinking wet sand. Flattery, threats and emotional blackmail had forced everyone from the high-ups in the federal police to the soldiers at Khazir checkpoint to submit to the doctor's will.

But the commander remained remote, furious and immovable. Twice, now, his people had sent messages to Dr Amir, reminding him what the two bullets were for. The doctor had thrown them into the grass near the checkpoint when he returned from their meeting, and hoped that they were now rusting away.

He could take it personally. Or he could get even. He decided on the second option and began to plot. There was only, the doctor reasoned, one realistic option left. They would trick their way past in disguise, hoping that the friendly soldiers on the checkpoint wouldn't sound the alarm if they somehow recognized them. They ran the risk of being captured and having the animals taken from them.

But for Dr Amir, it was the only option. 'At this point, I feel like they can just go ahead and shoot me,' he said to Marlies. 'Who cares. Let's do it.'

Dr Amir had amassed mounds of paper before he realized that the permits the bureaucrats and soldiers kept asking for were worthless as long as the commander said no. Bureaucracy had failed them, and the time for action had arrived.

The previous evening, Dr Amir had left the animals behind for the first time to go to Erbil. In the smoky lobby of the Classy Hotel, the team had gathered. All of them had taken turns spending days at the checkpoint, and were running on coffee and adrenaline.

Sitting at a table with the others crowded around him, Dr Amir explained his plan.

'Vegetables?' asked Yavor.

'Vegetables,' confirmed the doctor, grinning through his sleep-heavy eyes.

They called the office in Vienna, and the Vegetable Plan began to take shape.

At 12 noon the next day a truck would leave Erbil for Mosul. It would go directly to the zoo, passing by Dr Amir at the checkpoint without so much as a wave. An hour later, the doctor would give his great farewell and leave for Mosul to drop the animals off at the zoo for the last time.

In Mosul, after loading the animals off Dr Amir's truck and into the truck that had come from Erbil, they would stack boxes of vegetables up against the outside of the cages, so that when the doors opened the van appeared to be packed with cauliflowers and tomatoes.

Then they would cross the border and make their escape to Erbil.

It was a brilliant plan. Unfortunately, it did not work.

At around noon on the day of the planned escape, Dr Amir was Skyping with a journalist in the truck when he heard a commotion outside. Yavor came up to the window and told him there was a problem.

'I'm here with a truck for the animals,' shouted a voice in halting Arabic nearby. 'We're taking them to Erbil.'

Dr Amir threw himself out towards the truck that had just pulled in. This was the truck that had been meant to go straight into Mosul without saying a word about the animals. Instead, the driver was parked at Khazir. The guards, sleepy in their post-lunch haze, were ignoring him.

'What the hell are you doing?' said the doctor, grabbing the man by his lapels and pushing him on to his truck. He was shaking with anger The eight days of careful planning were about to be ruined. 'Are you stupid?' he hissed. 'Didn't I tell you exactly what to do?'

The man didn't reply.

Despite his anger, the doctor realized the driver was terrified, and let go of him. 'Go,' he said, and pushed the man back towards the driver's cab. 'Leave now. Go back to Erbil.'

He watched the truck drive away. The guards were staring at him.

Dr Amir's usually gentle disposition returned. 'Nothing to worry about,' he shouted cheerily. 'He's gone now.'

Back in the truck, he punched wildly into his phone. They would need to send a new truck from Erbil that day, he wrote. A different colour, so it wouldn't arouse suspicion. This time – he spelled it out – it would not stop at the checkpoint, but continue straight to Mosul.

No problem, wrote the truck company. The new driver would come soon.

Lunchtime wore into the afternoon, and the sun began to soften. Yavor, Gregor and Dr Amir sweated silently in the truck, waiting for the new truck to come. Marlies had gone back to Erbil to coordinate the vegetable delivery, as they had been referring to the plan.

Dr Amir had raised a stir across Iraq over the last few weeks, and made some of the country's most powerful men so angry they had threatened to kill him. It was reasonable to think that their communications were being monitored.

The afternoon sunlight was buttery gold by the time the new truck arrived. Dr Amir willed it to keep driving.

The lorry stopped by the soldier at the checkpoint. A moment later, it was rolling forward onto the road to Mosul.

Dr Amir turned to Yavor and Gregor. They all flashed

nervous grins at each other. It was happening. The Great Escape could begin. They pulled into the road, waving goodbye to the guards at the checkpoint, Lula and Zombie reclining in the back of their truck.

The original plan had called for them to wait an hour before following the other lorry, to allay suspicions that they were together. But there was no time. With the truck still in their sights, they started down the road to Mosul, 20 miles away.

Dusk was turning the horizon violet. Dr Amir sat worried in the cooling air. During the day, the road was relatively easy to navigate. There were three or four checkpoints, staffed by bored young soldiers from the Iraqi army and the odd militia. The war was still relatively far away in west Mosul, and though convoys of soldiers careered down the road every so often, it was safe enough.

Around 5 p.m. every day, that changed. The militias that lurked around the outskirts of Mosul – Shia groups, Christian fighters, Sunnis from the surrounding areas, semi-criminal gangs, all banned from the centre of the town – would drift in as the sun set and start setting up additional checkpoints on the road into the city.

The army set up their own extra controls, fearing the *inghimasi* attacks that came at night time – Isis sleeper cells regrouping behind army lines and emerging from their hideouts like rats from a drain to blow themselves up among the troops.

Each time Dr Amir's truck passed through one checkpoint, they would immediately see another one ahead.

It had taken them over an hour to go 5 miles. The soldiers were nervous and over-excited, and jumped whenever Dr Amir's truck approached, scattering around the checkpoint to shout at them to slow down, turn the lights on, say who the fuck they were. Then, the cry would go up: 'Lion! Bear! Let's do a selfie!'

They were battle-hardened men with guns and no interest in where the animals came from, or who owned them. They were also pathologically excited to update their Facebook profile pictures in order to impress their friends at home, and were in no rush as they posed over and over again to get the perfect shot with the lion. Dr Amir could have strangled them. Yavor and Gregor, though they were staying calm, jumped at the slightest noise. It was dark now, and they were still only halfway to the zoo.

'Where are you taking them?' the soldiers at the next checkpoint asked, yet again.

'To the zoo in Mosul,' said Dr Amir.

'There's a zoo?' the soldiers shouted, and the entire conversation would start over. The pieces of paper that Dr Amir showed them, all in bureaucratic English that they couldn't read, were glanced at, nodded at authoritatively and given back with an approving smile.

It was evening when they arrived at the zoo. Yavor and Gregor had stopped speaking completely. Dr Amir felt sick with the anticipation. Everyone knew this area was still at risk of attack by Isis sleeper cells.

Abu Laith was waiting outside the zoo. The zookeeper had taken to the new escape plan with untrammelled delight. When Dr Amir had told him on the phone, he had tapped himself under the eye, beaming with pleasure at the sneakiness of it all. Working quickly, he had assembled a trusted gang of children, neighbours and interested parties to help in the changeover.

They had been waiting outside the zoo for hours by the time the animals arrived. The others were tired, but Abu Laith was as excited as he had been at dawn that morning, when he woke up to start his preparations. 'Doctor,' he cried, as the veterinarian stepped warily out of the truck. The streets around the zoo were quiet, and Abu Laith's voice was, as usual, very loud.

'We've got everything you asked for,' crowed Abu Laith. 'Don't worry for a second. Everything is ready.'

Dr Amir thanked him and faced the others. 'We need to move,' he said. 'We've got ten minutes. Please can you try to keep things quiet?'

Zombie and Lula barely stirred as their cages were cranked from the truck onto the street outside the zoo. They had become used to the endless driving and being shifted about. Dr Amir had the strangest feeling that they somehow understood what was happening. They know we're on their side, he thought.

Abu Laith wasn't at all sure that he wanted the animals to leave again. The excitement at being part of the escape operation had started to wear off, and he was feeling decidedly worried. Somehow, though, he didn't quite believe the animals would make it out. Maybe the animals would be back soon, as they had been before.

'Let's go doctor,' he said, when the cages were resting on the ground. 'They're ready.'

In the thick darkness, Dr Amir and his crew drove away for what they hoped would be the last time. They would leave the last stage of the operation with Abu Laith. At 4 a.m., when no one was watching, he and his accomplices would move the cages onto a new truck. Then they would stack boxes of vegetables in the back to hide the animals, and a local driver would take them to Khazir and, they hoped, to freedom.

If anyone opened the back doors of the van, all they would see would be crates of vegetables, the green tops of the carrots peeking out. But if they lifted one of them away, they would see a fully-grown lion and a bear taking a nap in the dark interior.

It was late by the time Dr Amir saw the Khazir checkpoint appear in front of them, its bright lights flooding the ground.

'Assalamu aleikum,' cried the doctor, in his best attempt at cheeriness, as they rolled up to the checkpoint.

'Wa aleikum assalam,' the soldiers shouted back. They looked extremely concerned. Dr Amir realized they felt sorry for him.

'I'm sorry it had to end this way,' the guard said.

'It was impossible,' said his friend. 'Next time you'll manage it. You just need to get more permissions.'

Dr Amir nodded gravely. 'Absolutely,' he said. 'Next time.'

With emotional promises to be in touch, the doctor's car passed into Kurdish territory.

At dawn the next day, back at the zoo, Abu Laith watched the truck containing the vegetables, the lion and the bear drive off. It would wait down the road for a few hours before making for Khazir, to ensure that the truck wouldn't attract attention from anyone watching the zoo.

Dr Amir, who had been up since 4 a.m. drinking coffee and panicking silently, had left the hotel in Erbil and by 8 a.m. was sitting in no-man's-land near the border with Yavor, Gregor and Marlies, waiting for the car to arrive, drinking a can of cold coffee.

By 11 a.m., nerves had frayed to the point that the doctor, usually the calmest of all of them, couldn't listen to his thoughts any more. He turned the radio on, and Frank Sinatra's voice swept out of the open car doors.

Dr Amir sang along with him in his deep baritone. 'I did it my way,' he crooned, as the others gaped at him. They were almost jogging on the spot with anticipation, and were perplexed by the karaoke session. But as he sang, Dr Amir felt the fear lift from his chest, if only for a moment.

Then he saw a white truck in the distance, rolling slowly towards the checkpoint on the Iraqi side of the border. For a moment, he wasn't sure if it was the right one. Then he was, and his stomach roiled with fear.

This was the final hurdle. The Kurdish soldiers knew they were coming. There would be no problem there. All that remained was this Iraqi checkpoint that had thwarted them for nine days.

The truck rolled up to the checkpoint and stopped.

The doctor couldn't see what the driver was doing. No one spoke. He couldn't think. They all stared intently at the soldiers who stood around the truck in the distance.

The lion couldn't roar, the bear couldn't growl, the soldiers couldn't search the truck – or they would lose them in a moment. Dr Amir stopped breathing. Then, almost imperceptibly, the truck rolled forward a little. A moment later, with a surge of relief among the onlookers, it drove through the checkpoint and into no-man's-land. It came towards them, the driver sitting stony faced.

They had done it.

52

Dr Amir

It wasn't until they were a few miles into Kurdish territory that Dr Amir dared to stop their convoy and look at the animals. The group had decided to avoid the main road that ran straight to Erbil in favour of a multi-pronged dirt track that trailed away through the mountains. It would take them twice as long to get to the city, but it meant they avoided the checkpoints on the Erbil road.

Though they were out of Iraqi territory, they were still not safe. The commander in Mosul had men everywhere, and Dr Amir suspected that Ibrahim, however much he had professed his desire to help, would not hesitate to squeeze as much advantage as he could from their situation.

Dr Amir opened the back doors of the van and pulled out the top vegetable box, then the second. Through the gap he could see the bars of the cages, and in the dark behind them, Lula and Zombie sitting quietly in their cages.

'You didn't roar,' the doctor later remembered thinking. 'You knew what to do.'

He put the vegetables back and shut the doors. Time was short. They had to move as quickly as they could to enact the rest of their plan before anyone realized the animals were no longer in Mosul. Abu Laith and Hakam would cover up the escape for as long as they could, but there were only so many ways so hide the absence of a lion and a bear, especially if they were the only animals of their kind in the city.

Their convoy streamed up the road towards Erbil, through the great plains and empty villages that stretched up to the mountain peaks in the north. When they reached Erbil, they sailed through the checkpoints and on to the private zoo where Dr Amir had borrowed the cages.

For the first time in what might have been their entire lives, Lula and Zombie walked out from their cages and into a fenced enclosure. It was only small, but it was much bigger than their former cages. The lion and the bear who had grown up in Mosul Zoo looked around them, bemused, taking tentative steps, moving further than they ever had before. It must be nice for them to stretch out, Dr Amir thought, as he watched them tear into their dinner.

They wouldn't be here long. If everything went well, this was only a foretaste of their real freedom. A lot had gone wrong so far, and it was likely it would do so again. With this in mind, Dr Amir made a call to the airport.

The first entrance to Erbil international airport lies about 2 miles from the beer shops and churches of Ankawa, the Christian quarter of Erbil where the Classy Hotel does its roaring trade. But that is only the first entrance. The security restrictions at the airport – which lies next to an American air base – are so tight that the departures lounge and the runway and the surgically lit cafe selling pizza are a mile further on from that. There are two, sometimes three stages of security screening.

Two days after Zombie and Lula had made it to Erbil, Dr Amir, Yavor and Gregor were standing in the departures lounge, after finally making it through security. The day before, Marlies and the others had flown out from Erbil. Only the three of them remained with the lion and the bear for the final, perhaps most treacherous, stage of the rescue operation.

Zombie and Lula were back in their cages, in the cargo section of the airport. The men had left them there to go through security, where they had been told they needed to pay an extra $1,200 to get through passport control. Their cargo was overweight, it was claimed.

This was a problem. That morning, the doctor had maxed out his credit card paying the group's hotel bill. Ankawa had only a few ATMs, and most had an upper limit of around $500 on transactions. After weeks in Iraq, where very few businesses took credit cards, everyone was out of cash.

But they needed $1,200. At the doctor's request, everyone turned out their wallets. That raised only a few hundred. Dr Amir's pulse was rising. They only had a few hours before they were due to leave, and there wouldn't be another chance.

'Is there no way we can transfer it later?' the doctor asked one of the airport employees, again.

'No,' he said. 'I told you. You don't pay, you don't fly.'

It was all too much for Yavor and Gregor, who looked like they were about to cry. Weeks of work, blocked at every juncture. Now it was happening again.

In a state of utter exhaustion, Dr Amir contemplated breaking down, but decided it wouldn't help much. Instead, he began calling his way through his phone book, asking for help.

Twenty minutes later, Dr Amir put his passport down on

the counter and held his breath in anticipation. A colleague had come through with the money, and convinced the airline to let them on the plane. Through sheer luck and force of will, they might make it on their flight.

But as he sweated at the passport control desk, Amir thought of how many senior Iraqi officials had threatened him. One of them could have very easily added his passport to a blacklist. He was about to find out how serious their threats had been.

The officer scanned the passenger documents in front of him. Dr Amir tried to look calm, which was harder than it usually was, and smile, which was even harder. The officer seemed to be taking an unnaturally long time to check his passport and the accompanying documents from Four Paws. He was pausing at every page, flicking through the stamps and peering at the dates.

Then, with a final glance at Dr Amir, the officer stamped his passport. He was through.

With a low, humming screech, the cargo plane taxied up the runway. The doctor, still soaked with sweat after pushing the cages out of the lorry, looked out of the window at the retreating trucks that had brought Lula and Zombie into the belly of the plane along with at least ten armoured cars, which were the only other cargo.

A dose of tranquillizers in their food before the flight had left the lion and the bear sleepy and content. They had seemed happy enough when they went in.

Now the plane was speeding up, and the front wheels lifted off the tarmac. With a last glance around for pursuing vehicles, Dr Amir picked up his phone and wrote a quick message to the commander in Mosul.

'Thank you for your kind cooperation,' he wrote. 'We managed it anyway.'

He pressed send.

With a satisfied sigh, the doctor leaned back into his seat, and closed his eyes for what felt like the first time in years.

Epilogue

THE GRASSLAND OF SOUTH AFRICA'S FREE STATE STRETCHED out before the group as the cage was lifted off the back of the pick-up truck. Working together, the rangers heaved the cage on to the plain. The sun was high in the endless heavens, and ahead the trees shaded the ground.

Zombie was awake. Moving his shaggy head from side to side, he murmured gently – a thick, satisfied sound.

It had been three months since Zombie and Lula took off from Erbil airport. They had gone to Jordan, where they had been treated in an Amman zoo, under Dr Amir's watchful and exacting eye, until they were returned to full health. Lula's coat, previously grey and thin as a goat's, bristled with the lustre of ermine. At first, she had started and hidden from loud noises. Sometimes, Dr Amir knew, she still looked for her child. But the wounds of the past were healing. She played with the keepers, and ate enough for three.

A week previously, she had been released into a nature sanctuary in Jordan – many miles of fields, rivers and rocks to run and play among.

At Abu Laith's insistence, and the doctor's suggestion, Zombie was going back to Africa, where his parents had been bred or stolen from. The flight had landed in Johannesburg that morning, and Zombie had yawned as his cage was moved on to a truck. His mane was a rich orange now, rather than matte sienna. Muscles, gained from running in his enclosure in Amman, rippled beneath his skin as he examined his surroundings.

He had sat in his cage on the ride to Lion's Rock, perched on the back of a truck, the great open expanse stretching out around him. Dr Amir had helped open the sanctuary in 2006, in response to a wave of zoo closures in the Eastern bloc as governments scrambled to meet EU accession targets for animal welfare. Amid the crags and bushland of the Free State there lived over 100 big cats including lions, leopards and tigers plucked from caged lives of misery and given their freedom.

Now the workers were pulling up the bars of Zombie's cage. Sniffing, confused, the lion stood up and walked, experimentally, forward. As he stepped out of the cage, the rangers closed the door behind him.

Zombie looked up, and the sheer vastness of the plains unfolded in front of him, twitching with life and possibility. He bounded forward through the grass.

The lion from Mosul Zoo was finally home.

Postscript

It will not have escaped anyone's attention that while the animals of Mosul Zoo were rescued, the humans who kept them alive were not. Their lives have not been easy. They have struggled, just like the other civilians who survived the horror of Isis occupation and the battle to liberate the city. But they consider themselves to be luckier than others. They are alive, and they have their homes. At the time of writing, there are still bodies trapped under the rubble in the Old City, and thousands live in displacement camps after having lost their homes.

Abu Laith's hair has turned bright white, but he still strides around the neighbourhood, lecturing passers-by on his favoured subjects of hypocritical mullahs, spendthrift relatives and the general excellence of dogs. He is largely recovered after a bout of tuberculosis, which swept through Mosul in the aftermath of the liberation. Now he's back on his feet, he is consumed with plans to start a new zoo – bigger and better than ever. Lumia, who had another baby – Mustafa – after Shuja, is less than impressed.

Marwan still comes by Abu Laith's house occasionally, checking in on his old benefactor and bragging about his various conquests. He looked for Heba after the liberation, but never found her. He hopes she is all right. Luay is back at university, studying geography, though he still spends a large part of his time playing games on his phone. Lubna, Mohammed and Oula are back in Mosul after their years-long Baghdad sojourn. The children, as ever, roam about the house like a pack of small wolves, dragging their pets along with them.

Hakam started a band, Awtar Nergal, and is currently touring the world, most recently in Belgium, playing guitar. Not long after Isis left, they performed in the ruins of the Old City to a crowd of astonished Mosulawis. As if that wasn't enough, he's also back at the lab, working on his PhD. Hasna has got her first job as an English teacher, and spends her spare time reading more books than ever before. Said is back terrifying the criminals of Mosul in court, while Arwa is curating their garden; a paradise behind the peach courtyard walls. Their books are all back on their shelves, and new additions to the library have been made. The bathroom where the family hid for so long is now just a bathroom again.

Dr Amir had barely left Mosul before he was planning his next mission – a rescue operation to save the animals stranded in Aleppo Zoo. It came off not long after Zombie was taken to Africa. The doctor, who rarely stops, is currently planning another mission to Yemen. Marlies, and all of his other colleagues, are still by his side.

Lula lives on an enormous nature reserve in Jordan among other rescued bears. On the South African plains, Zombie runs free.

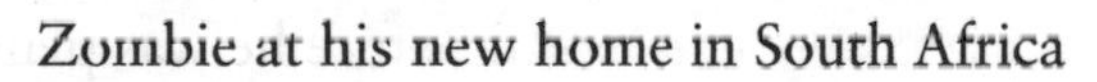

Zombie at his new home in South Africa

Sources and [illegible]

THIS BOOK DRAWS O[illegible]
[illegible] key people who [illegible]
[illegible] of them were [illegible]
subjects themselves. [illegible]
Where it wasn't possib[illegible]
– such as Heba, Mar[illegible]
memories of those who [illegible]

[illegible] and Abu [illegible]

[illegible] this book would [illegible]
[illegible] endless [illegible]
remarkable ability to [illegible]
[illegible] and times. A [illegible]

Sources and acknowledgements

THIS BOOK DRAWS ON EXTENSIVE INTERVIEWS WITH MANY of the key people who lived through these remarkable events. Most of them were conducted in Mosul in 2018 with the subjects themselves. The story is based on their recollections. Where it wasn't possible to speak to the subjects themselves – such as Heba, Marwan's girlfriend – I've relied on the memories of those who spoke to them.

In the case of the Commander, the reconstruction of events is based on Dr Amir's recollections. For the parts of the story relating to Ibrahim, the animal's former owner, I've relied on Dr Amir and Abu Laith's testimonies.

In a few cases, names have been changed to protect the person's identity.

As for the lion's name, Abu Laith calls him Zombie because he watched *The Lion King* with mis-translated Arabic subtitles. Four Paws calls him Simba.

This book would never have been written without Sangar Khaleel's endless supply of patience, good humour and remarkable ability to make distracted zookeepers remember dates and times. A lifetime of Fairuz-free mornings for you.

Thanks to Hakam, Hasna and all the Zarari family for their generosity with their time, cooking and gardening knowledge.

Thanks to Dr Amir, Marlies and everyone at Four Paws for their kindness, empathy and for putting up with my endless questions.

Thanks to Marwan, and to all of Abu Laith's children and their assorted friends, for remembering what happened, and for re-living it with me.

Thanks to Lumia, for her photographic memory and knack for recalling dialogue.

Enormous thanks to my editor, Neil Belton at Head of Zeus and – in the US – Diana Gill at Forge, for getting this book down to fighting weight, and making it so much better. Thank you too to the publicists, marketers, sales people and everyone else who worked to get this into people's hands.

To my agent Max Edwards, who interrupted a rambling story about some nice people I'd met at a zoo with: 'I'd read that.' Max, Lisa Gallagher in the US, and Samar Hammam in other languages, all fought for this book. Thank you.

Thank you also to my editors at *The Sunday Times*, especially Bob Tyrer, for giving me time off to write, and to Eleanor Mills for running an extract in *The Sunday Times magazine*.

Thanks to everyone who read early versions of this book and came with vital changes: John Beck, Sarah Dadouch, Stephanie Allen, you are all legends.

Thank you to Carolina Aguirre, who understood the book like no one else, and drew the original illustration from which the cover art is sourced.

Thank you to Mohammed Rasool, who sat through untold hours listening to complaints about greedy family members and still found time to be the world's foremost expert on Iraqi bread.

Thank you to Dr Anjam Ibrahim Rowandizy in Erbil, who saved Abu Laith.

Thanks to Salam Zeidan and Safwan Al-Madany, some of the kindest, funniest and most selfless people I've ever met.

Finally, of course, thanks to Imad Sabah, Abu Laith, a good man.

The world needs more people like all of you.

إلى أبو ليث وجميع أعضاء عائلته: كل الشكر على كرمكم وضيافتكم. من شرفي أنني تعرفت عليكم.

على راسي.

A note on Arabic nicknames

Abu Laith, of course, means Father of Lions. This is his *kunya*, a kind of Arabic nickname used by men and women across the region. Abu means father of, and is often followed by the name of a person's eldest son. For women, their kunyas start with Umm, which means mother of (Lumia is Umm Nour, while Abu Laith refers to his ex-wife Muna as Umm Laith).

It is, however, not always that simple. Abu Laith has had his kunya since he was a child – as many do. In his case it's due to his love for animals and shock of ginger hair that gives him a striking resemblance to a lion. When his first son was born, it felt reasonable to call him Laith.

Image credits

1. John Beck
2. John Beck
3. Ahmad Gharabli / Getty Images
4. Ahmad Gharabli / Getty Images
5. Getty Images
6. John Beck
7. John Beck
8. John Beck
9. John Beck
10. John Beck
11. John Beck
12. John Beck
13. Nur Photo / Getty Images
14. Mohammaed abu Qutai / Four Paws
15. Safin Hamed / Four Paws
16. Hakam Zarari

p. 375 Daniel Born / Four Paws

About the author

Louise Callaghan is the Middle East correspondent for the *Sunday Times*. She was named New Journalist of the Year in 2017, and won the Marie Colvin Award at the British Journalism Awards in 2018. The citation read, in part: 'Louise Callaghan's work fights to get to the truth of what is happening on the ground in rebel-held Syria... She bore witness to crimes governments and armed groups would rather were hidden away.' *Forbes Magazine* named her as one of their '30 under 30' key people in the media.